The Danube:
A River Basin in Transition

The GeoJournal Library

Volume 40

The titles published in this series are listed at the end of this volume.

The Danube: A River Basin in Transition

by

IRENE LYONS MURPHY

Colorado State University,
Fort Collins, Colorado, U.S.A.

SPRINGER-SCIENCE+BUSINESS MEDIA, B.V.

A C.I.P. Catalogue record for this book is available from the Library of Congress.

ISBN 978-94-010-6325-8 ISBN 978-94-011-5522-9 (eBook)
DOI 10.1007/978-94-011-5522-9

Printed on acid-free paper

To dear friends and colleagues in the Danube countries, allies in the struggle for peace, democracy, and a better environment, with gratitude

CONTENTS

PART FIVE: Long-term Commitments

Illustrations

Figures

Maps of River Basins

List of Tables

Preface

International river basins first captured my interest while I worked on water resource issues in the US Department of the Interior during the 1970s and '80s. I was especially intrigued with the way the US resolved long and recurring disputes with Mexico over the shared use of the Colorado River. The elements of decision making common to basins throughout the world were present in these transboundary conflicts and their resolution. In 1985 I began collaborating with Dr. Eleonora Sabadell, who was then a specialist in water resources at George Washington University and who is now with the US National Science Foundation, on analyses of several international river basin agreements. The work combined her expertise as a civil engineer and mine as a political scientist and practicing policy analyst. We wrote articles, participated in panel sessions, and in 1986 directed an international workshop at IIASA in Laxenburg, Austria.

By no means ready to retire in 1989 I nevertheless began to think of leaving Interior to spend full time exploring the challenge, and intrinsic value, of transboundary river basin agreements. I applied for a Fulbright research grant to develop for the Danube River Basin one of the most needed tools for successful management of an international river, a basin-wide information system. My proposal, for nine-months' study at the Bulgarian Academy of Sciences, was filed with the Fulbright Program on September 1, 1989. It suggested that since the information network would include Germany and Austria as well as the eastern bloc countries, it would serve as a "bridge between east and west." It said nothing about an anticipated coup in Sofia or in any other east European capital.

By the time I arrived at the Academy in Sofia in September 1990 a fire had shut down the offices of the Communist Party, the country's dictator was out of office, and arrangements were under way for free elections. I had intended to come home in June 1991 and write articles and/or a book about the Danube and then look for another river basin for travel and study. Instead I stayed on in Sofia to work on the first Danube-wide discussion of an environmental program for the basin and spent the next three and a half years on various projects to help implement it, focusing on two basic basin needs: data and information sharing and the strengthening of public participation in government decisions.

This book explores a decision framework for international basin management and applies it to the Danube. Measured by accomplishments elsewhere in the world the Danube experience with management has scored impressive achievements. Lessons learned can support the continued growth of democracy in eastern Europe and suggest management improvements for other basins throughout the world.

First of all I acknowledge with deep gratitude the help of the United States Institute of Peace which funded a 12-month study, "International Cooperation in the Danube River System, 1990-1995," and made the writing of this book possible. Special thanks go to Dr. Steven Riskin, the Institute program officer for the project. I am grateful as well for the assistance provided by the International School for Water Resources at Colorado State University and especially to Professor Evan Vlachos, its associate director. The fortuitous award of a NATO Advanced Research Workshop grant brought together specialists from all the Danube countries in May 1996 to review Danube data- and information-sharing. Their conclusions and recommendations are in the proceedings published in December 1996 and cited in the bibliography.

During my years working and living in the basin, first in Sofia and then for three years in Budapest, I received funding and other help which contributed to research for this book. Support was received from the U.S. Fulbright Program, the Bulgarian Academy of Sciences, the European Bank for Reconstruction and Development (EBRD), European Union Phare Program, the Regional Environmental Center (REC) in Budapest, VITUKI, the Hungarian water resources research institute, and the UN Development Program. The program coordination unit of the Environmental Program for the Danube River Basin and its team leaders, first David Rodda and then Teun Botterweg, and Richard Holland, formerly with the EC and the EPDRB staff, have been generous with copies of documents and other help as has Hellmut Fleckseder of the secretariat of the interim International Commission for the Protection of the Danube River.

From the start east European specialists in water resources, in governments, research institutes, and nongovernmental organizations, have provided expert help of all kinds and given me data and information agreeably and promptly. The knowledge and good will of all was appreciated but especially that of Delcho Vitchev, first in Bulgaria, then at REC, and then at EBRD in London; Ilya Natchkov and Ivan Milushev of the Bulgarian Ministry of Environment, Borislav Georgiev and Miroslav Iliev of the Bulgarian Academy of Sciences, and Atanas Paskalev of the Bulgarian National Association on Water Quality; Bogdan Paranici of the Ecologist Youth of Romania (TER), partner in projects and an excellent guide to the Danube Delta among other natural areas; Maria Galambos, Ministry of Environment in Budapest; Peter Bakonyi, director of VITUKI and Kalman Morvay, formerly of VITUKI; Mitja Brilly, University of Ljubljana, and Aleksandra Kornhauser, International Center for Chemical Studies, Ljubljana, Slovenia, and Ivan Zavadsky, Ministry of Environment, Bratislava, Slovak Republic. I also want to thank Karin M. Krchnak for providing me with information about the transition in Slovakia as well as the Gabcikovo case before the Interntional Court of Justice; Bruce Campbell for help with the Danube country economic indicators; Hame Mamadou Watt, University of the District of Columbia, for information on the Senegal Basin, and William J. Barnds of Washington, DC, for improving the focus of this study and for general insights about international relations.

It is, of course, understood that errors and omissions are my responsibility alone.

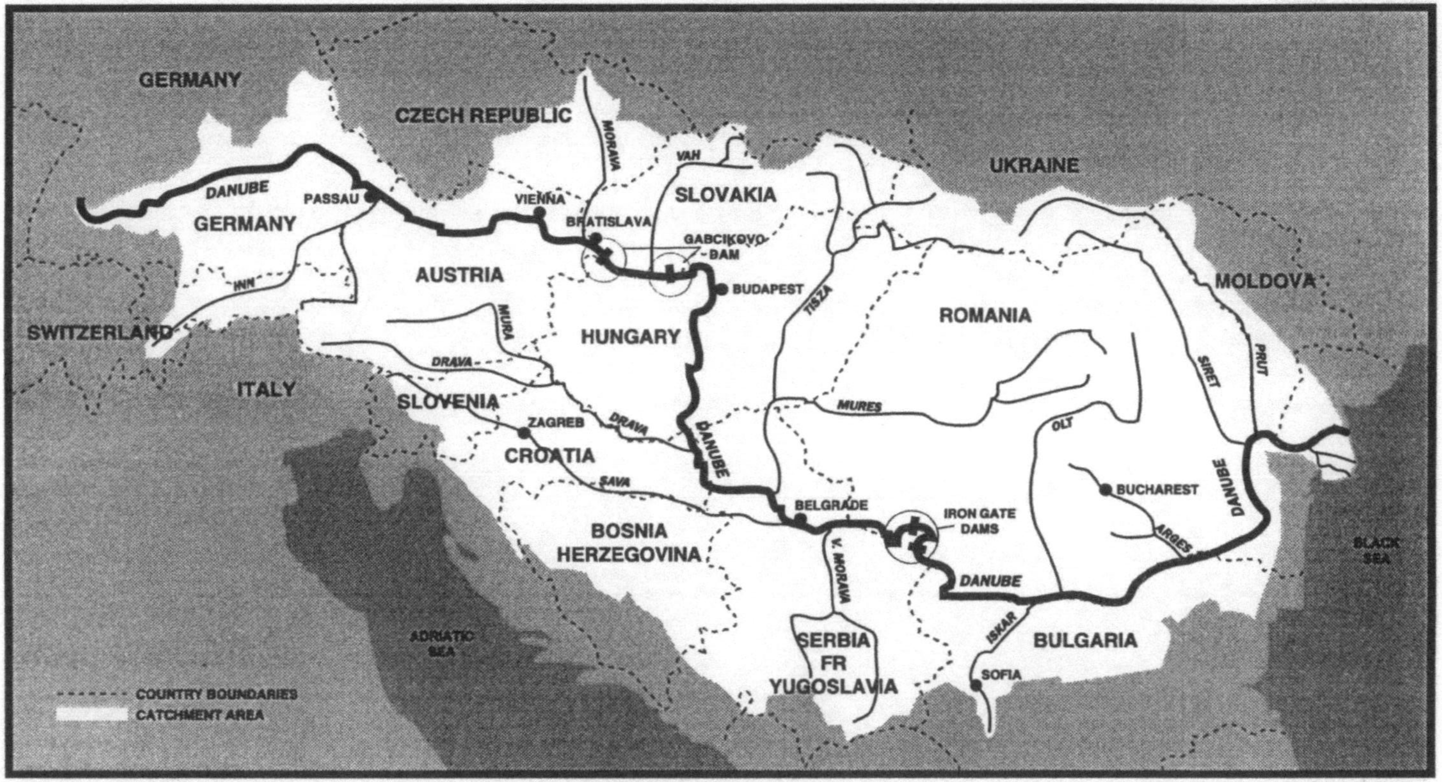

Danube Basin, 1996

CHAPTER 1

Danube Management in Transition: an Overview

1.1. Instruments of Peace

International river basin agreements can be instruments of peace in regions accustomed to conflict. Conventions can endow transboundary commissions with authority to carry out the collective decisions of riparian countries, making the river safe for navigation or from possible harm from dams or pollution. Basins can be managed by agreements that protect natural resources and offer all countries equitable shares in them. Needless to say there are far more river basins around the globe that need cooperative management than have it.

1.1.1. WAR

From 1991 to 1995 the instruments of war were ascendant in the former Yugoslavian Federation and its four Danube Basin republics: the riparian countries of Croatia and Serbia, and Slovenia and Bosnia-Herzegovina, with major Danube tributaries. Gray gun-boats invaded the wide, smooth reach of the river downstream from Budapest, violating the Danube navigation treaty and sacking the port city of Vukovar. The Sava, the Drava, and the mainstream Danube, damaged by eco-terrorism and the refuse of armed conflict, could no longer properly serve their basin populations. The settlement in Bosnia-Herzegovina may not mark the final phase of a conflict which has taken 300,000 lives and displaced at least two million people.

1.1.2. PEACE

And yet, during all this time representatives of eleven Danube countries were collaborating on the installation of a permanent instrument of peace in the basin. Slovenian and Croatian government experts came to meetings and workshops throughout the Danube region even while their countries were under attack from Serb tanks and artillery. They joined government and research teams to activate a program to clean up the Danube and its tributaries. Their work included the signing of a formal convention by all the cooperating countries, expected to be ratified in mid-1997. Its commission will include member countries and the EC; they will take part in significant measures to protect the natural resources of the basin. Bosnia has indicated its intention to join these activities. Serbia will be invited to join when it is ready.

The negotiation process and its product are unique in many respects. The convention embodies major principles of international river basin management developed in recent decades in response to increasing water scarcity and environmental damage. Agreement on the convention took place among countries isolated from western Europe for five decades and pre-occupied with transforming their ideologies, political institutions, and economies to match the norms of western democracy. International funding agencies, increasingly aware of the need to build environmental protection into economic development assistance, guided and funded collective plans for the rehabilitation of the basin's severely damaged resources.

1.2. River Basin Rationale

The principles of management now guiding crucial collaboration in the Danube Basin are grounded in international law and the history of recent basin development. River basins occupy almost fifty percent of the world's land area, providing resources for more than half the world's population. The scope of this study is limited to the policy decisions which key actors must make to achieve the benefits of agreement. Significant variables outside the policy process influence such decisions. They include economic and social factors and the day-to-day and long term management of the availability and use of natural resources. Both require the guidance of economists, water experts, hydraulic engineers, hydrologists, and specialists in other disciplines.

The success of the policy decision process depends on linkage with accurate expert sources. Decision outputs have a continuous impact on world health, economic development, and on local, national, and international conflicts. Peaceful negotiation of disputes over the use of rivers can lead to improved access to water for regions beset with the severe consequences of its increasing scarcity and can point the way to settlements of related disputes.

Successful agreements depend on three major accomplishments. Basic problems interfering with the protection and development of resources must be acknowledged, understood, and their solution sought by all basin countries. The goals of agreement must match the basic requirements of basin management. Finally, the crucial element of support for long-term management of basin resources comes from the countries themselves. The goals of each country's domestic and foreign policy must be mutually supportive over time to enable them to reach effective basin agreements at key decision points..

Navigation regulations are in place in major rivers because economies need transport for trade. Benefits over-ride international rivalries unless the countries are at war or close to it. Conventions to protect river resources from the transboundary impacts of development are rare, whether they are generated by diversions of the mainstream for hydroelectric power projects or by failure to provide adequate treatment for agricultural, municipal, or industrial waste. Lessons learned from the histories of five major river basins in four continents illuminate the rationale of basin management as it has been applied or mis-applied across national boundaries and help in the analysis of the Danube Basin (Chapters 2 and 3).

1.2.1. DANUBE ISSUES: PRE-TRANSITION

Historically, the Danube Basin appears to have had few static interludes. Centuries of conflict were less the result of battles for scarce water and more a question of ensuring a friendly buffer zone for competing power blocs, east and west. If international politics can be likened to a chess game then the players seeking to win power in the basin have retained the same goals and used essentially the same strategies for well over a millennium. The players, east and west, used the bishops and knights of religious, economic, and military persuasion to win control of the river and its resources. The pawns were the settlers from south and north who sought to gain sustenance and national identity along the banks of the Danube and its tributaries.

Acute conflicts between countries in the present-day Danube Basin can be traced to the early 20th century when the international order established by the Congress of Vienna was finally shattered (Chapter 5). Congress agreements had brought a general peace to all of Europe in a number of ways, including the affirmation of freedom of navigation on European transboundary rivers. The long-standing Danube River navigation agreement was re-negotiated twice following World Wars I and II. In 1947 the Soviet Union took the opportunity to exclude western powers from participation, confining membership in the Danube Commission which regulated navigation to the eastern bloc countries and itself.

Socialism, 1945 to 1989

The imposition of Soviet style governments and economies on eastern Europe after World War II was supposed to help Danube countries improve the use of joint basin resources (Chapter 6). The coordination promoted by Moscow, however, exacerbated industrial and agricultural pollution and caused a general deterioration in natural resources. Uniform environmental standards were included in laws copied from Soviet models but east European governments failed to enforce them. The Council for Mutual Economic Assistance (CMEA) generated economic policies which lessened opportunities for profitable use of local resources and mandated industrial specialization to serve Russian rather than east European country needs.

Each country reacted uniquely to Soviet domination. Some slavishly followed Soviet models (Bulgaria) while others declared varying levels of political and economic independence (Yugoslavia, Hungary, Romania). Growing awareness of the dangers of transboundary pollution in the 1960s and 1970s led to increased air and water quality monitoring but enforcement of emission levels from factories and municipal sewage systems was lax. In the late 1980s citizens mobilized by government failures to address life-and-death issues of air and water pollution contributed to the overturn of the communist governments, notably in Hungary and Bulgaria.

An understanding of the unresolved issues of the communist period is essential to the current search for the effective management of basin environmental issues, particularly as they relate to general political and economic development in the region (Chapter 7). The Cold War period offers an opportunity to look for patterns in domestic and foreign policies which continue to affect international cooperation in the Basin during the transition.

Limits to Danube Cooperation
The Rhine countries, shocked by the visible deterioration of their vital river system, responded in the 1950s and 1960s with joint decisions to repair damage and reduce the dumping of waste into tributaries and main stream. Danube cooperation was limited to navigation regulation through an international commission in which Germany had observer status and the official languages were French and Russian. Belatedly, the Danube countries gathered in the mid-1980s to agree to monitor the level of pollution at eleven river sites. Monitoring technology was incompatible and data collected were flawed. In any case, data were shared long after parameters had been measured and not used for policy purposes.

1.2.1. POST-1989 OPPORTUNITIES

The Danube east European countries joined the west in international agreements to reduce air and water pollution and preserve biodiversity. There was a substantial gap, however, between endorsement and vigorously protective national policies.

As western Europe followed or improved on the EC Council environment directives of the 1970s and 1980s, the state of the environment in the border region to the east was an increasing problem. The Chernobyl disaster in Ukraine raised concern to a significantly higher level. Carelessness in resource protection was one thing but failure to monitor and prevent nuclear reactor explosions was quite another. The post-1989 government changes provided an opportunity for western support for more effective environmental policies in countries in transition.

The Danube was an obvious target for a collective response to serious environmental problems. Post World-War II industrial and urban development clustered along its tributaries. Waste from munitions factories, chemical and metallurgical plants, pig farms and municipal sewers was dumped untreated into streams and into the main river itself. Disaster areas dotted the landscape from the deadly effects of burning soft coal in Czechoslovakia to the forsaken farming villages of the delta and rows of rotting fish on Black Sea beaches. Transboundary clean-up required national willingness on the part of all the Danube countries to take effective action, establishing and enforcing environmental standards compatible with those in western Europe. When the new governments in eastern Europe began to seek European Union membership environmental policies became a necessary item on transition agendas.

1.3. International Agreement on a Program

An informal team began discussing the possibilities of a Danube environmental program early in 1990 (Chapter 8). A core unit, with staff from the World Bank, the UN Development Programme (UNDP) and the Commission of the European Communities (EC), with the assistance of the U.S. Agency for International

Development (US AID), agreed to coordinate and channel aid to eastern Europe to benefit the protection of Danube River Basin natural resources. Two meetings in 1991 demonstrated the willingness of basin countries to cooperate with the international funding agencies to reach the newly established goal.

At the first meeting, held in Budapest in February, the possibility of a basin-wide convention which would enlist support for environmental protection was discussed by representatives of all the countries and the major funders (Chapter 8). At the second meeting in September representatives of the Danube governments met with international funding agencies to support the establishment of an Environmental Programme for the Danube River Basin (Danube Program). A task force of donor agencies and government representatives was formed to carry on cooperative activities pending the ratification of a formal convention and the establishment of an international commission. A detailed work plan was developed early in 1992 and a three-year program operating by the middle of the year.

1.4. The Transition in Danube Countries

The Danube Program began its environmental cleanup campaign in 1992 in relative obscurity. The opening of two major projects on the Danube that year, a hydroelectric facility built by Slovakia at Gabcikovo and the connection of the Rhine and the Danube via a 106-mile canal, received more attention, partly because they both were perceived as serious threats to basin resources. Other potentially significant conflicts were in evidence. Diversion of the river and its transboundary tributaries had changed their courses and limited their use by downstream countries. Variations in the level of political and economic development in basin countries were evident everywhere.

"Third parties," occasionally a euphemism for international agencies allied with large consultant firms seeking opportunities to build large dams, clearly had been and might still be important for environmental management successes and failures. For the first time since World War II international economic sanctions and armed conflict severely crippled the use of the river for transport inflicting uncalculated but severe damage to economies and natural resources in the lower basin.

The governments of participating Danube eastern bloc countries gave support to a detailed set of joint Danube pro-environment activities early in 1992. The primary goal, compatibility with western environmental policies, depended on enacting and enforcing extensive legislation. New monitoring techniques had to be installed and environmental data freely exchanged. Data were needed to evaluate the most severely polluted sites and determine treatment. Funds had to be allocated for all these purposes.

The willingness and ability to make extensive changes in environmental policies varied from country to country (Part Four). Adoption of western standards was made part of the criteria for EU membership for six of the former eastern bloc countries. All were applicants except for Croatia, Bosnia, and the Serbian Republic of Yugoslavia. The two CIS countries, Moldova and Ukraine, had pledged their participation in the Danube Program and taken preliminary steps toward political and economic reform.

Brief summaries of the status of political and economic reform are followed by more detailed descriptions of the environmental problems inherited from the past and the measures that have been taken to resolve them. Like political change, measurement of environmental policy development cannot be charted statistically. In any case much environmental quality data from the pre-transition period are unreliable and cannot be used for comparative purposes. The countries have been helped to provide accurate and mutually compatible data and information on pollution sources (industrial and municipal discharges, agriculture, among others) and the type of treatment they need.

1.5. Danube Achievements

The Danube Environmental Program (Part Five) was intended as a bridge to support the future activities of a new International Commission (IC), the product of a basin-wide convention. Despite major constraints, by 1995 basin countries had:

- agreed to solve problems in a common forum and to develop and share environmental and other essential data freely with each other;
- prepared, ratified, and entered into force a basin-wide convention which provides long-term assurance for the continued restoration and protection of basin water resources through a Commission, which would serve as a forum for future basin-wide decisions and their implementation;
- contributed to a basin-wide strategic action plan to restore and protect water and other natural resources, and
- made progress toward achievement of their goals with the active cooperation, including funding, of investors committed to respect the collective plan for restoring and protecting environmental resources.

Their achievements were not a precise match of the model of international river basin management described in Chapter 2, but they were impressively close.

1.5.1. THE CONVENTION

The Danube Convention which was adopted by the Danube countries in 1994 and expected to enter into force in mid-1997 provided a long-term mechanism for strengthening and implementing the joint environmental policies begun in 1991 (Chapter 21). The convention supports the harmonization of national environmental policies including the adoption of principles considered to be essential for effective enforcement. It affirms the need for the development and effective sharing of accurate, timely water quality data and information throughout the basin. The convention is a detailed but compact document based on the cooperative principles of international river basin management expressed in the Helsinki Convention and other documents and on the unique needs of the Danube Basin.

The convention commits countries to establish and perpetuate the transfer of accurate, compatible data and information about the status of their natural resources (Chapter 22). The Danube Program initiated a comprehensive set of activities including an accident and emergency warning system as well as a network of sites in each country to monitor, record, and share data about water quality. A NATO workshop of country representatives and river basin specialists made recommendations for improved access to water quality data and information and promised to follow-up with electronic prototypes for future adoption by the IC (Murphy, 1996).

1.5.2. INSTITUTIONAL STRENGTHENING, REHABILITATION PLANNING

The need for new "civic societies" in the former communist countries was widely recognized. Institutions to be reformed along with governments and economies included the political party system, the media, and non-governmental interest groups. Limited help was provided in this sector (Chapter 23). More help to grass roots organizations, including the provision of coordinated national services to them, was promised.

The founding session of the Danube Program in Sofia in 1991 recognized the need for a basin-wide plan (Chapter 24) which would prioritize environmental protection goals and specify the national and international decisions necessary to achieve them. The Strategic Action Plan (Plan) devised in 1993 and endorsed by all Danube countries in 1994 described over 170 Basin "hot spots" in need of new or improved treatment facilities and others requiring biodiversity support. A more specific implementation program and national action plans set more specific goals for environmental policies including the construction of waste water treatment facilities and suggestions for their financing.

1.5.3. THE INTERNATIONAL COMMISSION TAKES OVER

At the end of 1995 the Danube Program Task Force could count significant accomplishments. Two problems needed to be resolved. The transfer of operational responsibilities from the Task Force and its coordinating unit to the International Commission (IC) and its secretariat was one and finding funding for the future support of the Commission and the basin-wide action program was the other. The latter appeared to be the more difficult problem.

Arrangements were made for an interim secretariat to join the PCU at the latter's headquarters in the International Vienna Center and plans established for an orderly transfer of responsibilities to it. Commitments from both the EU Phare and GEF programs to support a second phase for the PCU and the Task Force guaranteed funding until the establishment of a permanent Commission. The very large investments needed for waste treatment and other facilities to ensure resource protection management had been generally identified.

The EU Phare program estimated the total cost for all of eastern Europe for these purposes at 300 billion ECU. Proposals for funding Danube Basin needs included a trust fund and a possible "Marshall Plan" approach, the Danube Recovery Program, to be presented to a Ministerial Conference in 1997.

1.6. Recommendations, Conclusions

It would be safe to wager substantial sums of money that the carnage initiated by Serbia in post-Cold War Yugoslavia and its potential dangers for eastern Europe will generate considerably more commentary than the valiant, under-funded, necessary struggle to contain pollution in the Danube Basin that was waged at the same time. Workshop meetings to discuss new legislative standards, improved monitoring, and investments for waste treatment plants were held close to war zones. The hypothesis that the new International Commission to Protect the Danube River might serve as an instrument of peace in a region prone to violence has not yet been tested. Firmly entrenched, the agreement to protect the vital resources of the Danube and to guarantee their equitable use across borders could improve the health and well-being of a now endangered population, and at the same time demonstrate the viability of peaceful settlement of regional conflicts (Chapter 26).

PART ONE: Managing River Basins: Benefits and Challenges

CHAPTER 2

A Rationale for Basin Management

2.1. Coordinating Aid

Aid to the new democracies in eastern Europe after the collapse of communism, it was generally agreed, should not be scattered piecemeal but rather contribute to systemic reform in critical sectors, "focusing maximum attention on those areas (Hutchings, 1994, p. 176)." Aid for political and economic reform eventually emphasized the need for eastern recipients to meet western environmental standards as funding for new waste treatment infrastructures took a higher place on the western aid agenda. But specialists in the EU Phare (originally "Poland and Hungary: Aid for the Restructuring of Economies," eventually aid for all Central and European countries) program and the international financial institutions, particularly the World Bank, suggested that aid could follow a framework for improved basin-wide management, pending agreement among countries and donors.

The nearby Rhine was a case in point. International commitments among basin countries helped to restore vitality to a dying river. Lessons had also been learned throughout developing regions of the world. Disruptive and costly conflicts in these regions had resulted from a lack of commitments to the protection and equitable sharing of resources. A model for wise management emerged in the 1980s as critical water shortages threatened and development conflicts grew in intensity. The guiding principles needed to be understood and applied rigorously if optimal solutions to river basin conflicts were to prevail.

2.2. A Rationale for Management

International river basin conflicts have been the subject of negotiations for decades. Such conflicts, regardless of type or intensity, are grounded in the condition of basin natural resources and the economic and political history of each country. While basins with the most serious problems are in the developing parts of the world, conflicts exist everywhere.

2.2.1. ROOT CAUSES OF CONFLICT

The identification of four typical sources of conflict is the first step toward management agreements:

- the impact of geographic setting and hydrogeological features on the use of basin n natural resources;
- variations in the level of political and economic development in each basin country and their influence on basin resource use;
- the effect of investors and other outside interests on development of river resources;
- historic national and regional rivalries and the way in which they have (or have not) been resolved.

A number of uses conflict with each other and get in the way of equitable resource sharing. Conversion of international rivers and their tributaries into waste disposal sites for cities and industries contributes to tensions among basin countries. A large part of the world's population is threatened by increasing water shortages. Countries in the developing regions of the world will experience almost ninety per cent of the doubling of the world population expected to occur somewhat after the middle of the next century. Environmental damage to resources, already severe, will also increase as industrial and other kinds of non-agricultural development deplete supplies of clean water. There will be a delay in water resource project construction because of escalating cost, the lack of investment funds, and the increasing technical complexities of new projects.

The possibility that international river basin conflicts may lead to war is frequently raised. Basin resources may become "both the targets and the tools of war." Peter H. Gleick reminds us that historical targets of war have included the water supply canals of Babylon in the seventh century BC, whose demolition guaranteed victory to the attacking Assyrians (Gleick, 1993, p. 87). The control of water systems through direct attack or sabotage has been a war strategy for centuries. Hydroelectric dams were targets during World War II and, more recently, during the Persian Gulf War. In January 1993 explosives planted intricately and extensively in Croatia's Peruca Dam destroyed a regulating tower and released millions of cubic meters of water into the river valley below. The dam, in a cease-fire zone, will have to be completely rebuilt (*World Rivers Review*,8:1(1993), p.1).

2.2.2. BASIC MANAGEMENT SUPPORT

A meeting of international river basin experts held only a few kilometers from Vienna at the International Institute for Applied Systems Analysis in Laxenburg in 1986 drew on these lessons to reach conclusions about basic requirements for international river basin management. The year was 1986; the nearby Danube was not on the agenda. It seemed out of the question that the two upstream western countries of Germany and Austria and the downstream countries of Czechoslovakia, Yugoslavia, Hungary, Romania, Bulgaria and the Soviet Union would agree to common pollution control policies. During this time hydrologic data which might aid navigation was shared but access to accurate data about levels of nitrates, arsenic, fecal matter, among other contaminants, was not available to the west or, in many cases, within the eastern bloc countries themselves.

The participants in the IIASA workshop discussed decision processes within four major international river basins: the Senegal, the Mekong, the Indus, and the De la Plata. Water specialists, members of river basin commissions, and staff members of international agencies, described the basic requirements for international river basin management (Murphy, 1987). All basin countries must

- solve problems in a common forum and develop and share accurate environmental and other essential data freely with each other;
- commit to a basin-wide convention which provides long-term assurance of an equitable use of basin resources and delegates an adequate measure of authority to an implementing agency;
- agree on a basin-wide economic development plan which combines optimal use of natural resources with environmental protection, and
- have the cooperation of international banks and other investors in the implementation of the development plan.

2.3. River Basin Management in Practice

Support for management rationale suggested at the IIASA meeting has increased in recent years. Positive trends include greater participation by all basin countries in development decisions, the growing interest in and affirmation of relevant principles of international law, and stronger environmental policies in countries which have been made aware of the dangers of transboundary pollution. A complex set of variables guides decisions in each international basin, however, and the perception of gain from management agreement may be greater than its actual effect.

Resource scarcity has caused countries to use position on the river to divert its resources for their own use at the expense of other nations. Bi- and multi-lateral agreements are made to support such use; help may be sought from or be provided by third parties interested in basin development. Similar decision-making patterns are found in all international river basins. The increasing need for individual countries to maintain positions of economic and political strength in the basin deter the consummation of basinwide agreements which seek to ensure equitable use for all and have prevented such agreements in all but a handful of basins.

2.3.1. GLOBAL APPROACH EMPHASIZED

The need to reinforce regional and global solutions to natural resource conflicts is more and more widely recognized. At a UNEP-UNDP forum in Geneva in 1993, experts recognized that "[a]ll over the world the concern with water resources, increasing environmental awareness and a series of natural resources 'crises' have coalesced in reinforcing a trend toward comprehensive management and interdisciplinary planning (Vlachos, 1993)." The former director of the International Institute of Applied Systems Analysis, Robert H. Pry, states in the foreword of a book on international environmental negotiation that despite advances in some parts of the world, "[t]oday, the increasing toxicity of the air we breathe, the water we drink, and the earth we walk on is so critical to the lives and well-being of all of us, that environmental negotiation is nothing less than an exercise in negotiating the survival of society (Pry, 1993, vii)."

A series of international meetings in recent decades supported a global approach to long-term management of transboundary river systems. The UN Water Conference in Mar del Plata, Argentina, in March 1977 established an action plan which urged states sharing resources to cooperate in existing programs and procedures, and in coordinated development of water resources (Gleick, 1993, p. 435). Cooperating countries were also asked to help prevent water-borne diseases and reduce or mitigate drought, flooding, and other man-made and natural hazards. The plan provided detailed objectives for the institutional arrangements to achieve such cooperation, including the prevention and settlement of disputes.

Subsequently the UN General assembly declared the period of 1981-1990 to be the International Drinking Water Supply and Sanitation Decade. While widespread activities sought to improve water services, especially in developing areas of the world, little overall progress has been reported. In sub-Saharan Africa (see details of the Nile and Senegal river basins in Chapter 3), for example, the number of people lacking access to clean water increased by almost thirty per cent during this period; despite the fact that the actual number of people receiving services had doubled (Biswas, 1992, p.71).

2.3.2. APPLYING INTERNATIONAL LAW

The UN International Law Commission (ILC), established in 1947, has suggested changes in current international law or new law in response to the challenges of river basin conflicts. Countries have been slow to agree on the codification of such changes, access to clean water increased by almost thirty per cent during this period; despite the fact that the actual number of people receiving services had doubled (Biswas, 1992, p.71).

2.3.2. APPLYING INTERNATIONAL LAW

The UN International Law Commission (ILC), established in 1947, has suggested changes in current international law or new law in response to the challenges of river basin conflicts. Countries have been slow to agree on the codification of such changes, however. In general, legal obligations have followed principles agreed to by participants in the 1972 UN Stockholm Conference on the Human Environment, which established that a nation had a sovereign right to exploit national resources pursuant to its own environmental policies, but that it also had the responsibility to ensure that activities within its jurisdiction or control not cause damage to the environment of other states.

This concept was echoed at the Mar del Plata meeting and elsewhere. Major recognized principles of international law applicable to rivers include the concept of equitable utilization, which "means that each basin state is entitled to a reasonable and equitable share in the beneficial use of shared water..." to which Gleick adds the obligation to "notify and inform other nations of any activities on shared watercourses that will affect them." An exchange of water resource data should be widely encouraged (Gleick, 1993a, p. 106-108)."

2.3.3. UNECE CONVENTION

The UN Economic Commission for Europe (UN ECE) which has 22 member countries from east and west, including the United States, adopted a "Convention on the Protection and Use of Transboundary Watercourses and International Lakes" in March 1992. This re-enforced the need for regional bi- and multi-lateral treaties among riparian countries to control water pollution. In taking appropriate measures, the document concluded, "[w]ater resources shall be managed so that the needs of the present generation are met without compromising the ability of future generations to meet their own needs (UN ECE, 1992, p. 165)."

While such principles may be acknowledged, international river law is one of the most unsettled areas of international law. There are few rules and these are usually set forth in on factors outside the policy process. Climate, hydrology, resource contamination from outside sources, general economic conditions, and the impact of hostilities elsewhere must be considered be controlled. In developing countries, the policies of public and private sector investors play a powerful role, while the effectiveness of decisions made by basin countries is often limited.

A framework for the analysis of national and international decisions which affect the acceptance of management goals, therefore, must focus on the complex set of factors influencing such decisions. The following chapters, which provide a "macro" view of five basins in four continents, identifies more failures than successes in the resolution of their management conflicts. It suggests the importance of economic and political stakes in such disagreements, and evaluates the collective response to management needs. The review of five major international river basins is intended to lay the groundwork for a detailed evaluation of basin management decisions and contribute to the questions which must be raised and answered if the consequences of such decisions are to be understood.

CHAPTER 3

Four Basin Studies

3.1. Applying River Basin Management Rationale

Most international river systems are not managed so that all of their countries share equitably in basin resources. Failure to protect one or more countries from the harmful action of others in the basin can be costly. The results of failure are measurable in a number of ways. Lack of access to safe drinking water for basin countries frequently causes debilitating illnesses and high morbidity rates. Loss of water for irrigation results in malnutrition and famine. Drought due to natural or manmade causes reduces protection for wetlands, fish stock, and ground water. Conflicts among basin countries between "haves" and "have-nots" can result in loss of life and a failing ability to cope with resource scarcities. Weakened political and economic infrastructures become less able to deal with critical domestic needs.

Five basins altogether have been chosen for review: four in developing regions of the world and one in Europe, the Rhine, which borders and is linked with the Danube Basin (Chapter 4).

Four of the basins, the Nile, Senegal, Lower Mekong, and De la Plata, are in developing parts of the world and are especially subject to population and water-shortage pressures. All have experienced the effects of conflicts over chronically scarce resources, compounded by ethnic and ideological struggles. Immense population dislocations and loss of life have resulted. Millions of people have died within these basins in the last two decades, succumbing to war, famine, disease, and genocide.

These phenomena in most cases cannot be explained entirely by mismanagement of natural resources nor can we say that basin-wide agreement to improve equitable use of them would resolve conflicts deeply rooted in the past. And yet the failure to manage resources has contributed substantially to the serious problems besetting each basin. In any case, improvement in management must occur if resources are to be protected and potential catastrophes averted.

The possibility for conflict in the Nile basin has been noted by Boutros Boutros-Ghali, former foreign minister of Egypt and secretary general of the United Nations. He is quoted in the *Financial Times* (April 4, 1993) as saying that the "next war in the Middle East would not be over politics, but over the waters of the Nile." Such assertions of the possibility of conflict echoed the 1978 warnings of the late President

Egypt who said that if Egypt was deprived of its Nile water it would not hesitate to go to war.

3.2. The Nile: One-country Monopoly, Growing Scarcity

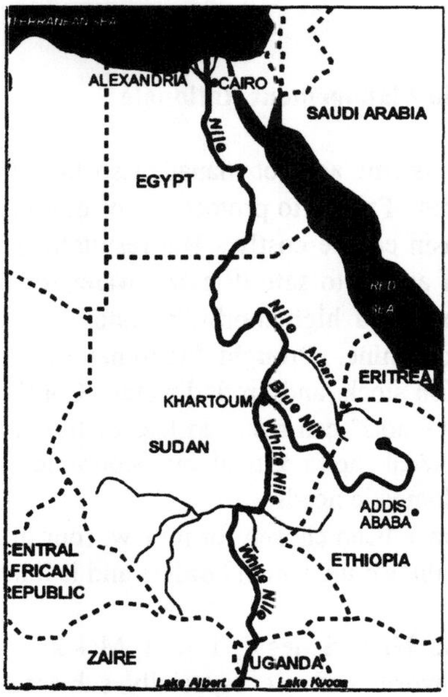

The Nile River Basin

Monopoly of the Nile by Egypt began five millennia ago. The world's longest river, it measures 6,632 kms. (4,145 mis.). Its major tributary, the Blue Nile, is 1,610 kms. (1,000 mis.) long and rises in Ethiopia, joining the mainstream "White" Nile at the Sudanese capital, Khartoum. All but 15% of the annual river flow into Egypt comes from the Blue Nile. The Nile and its tributaries drain an area of over three million square kilometers (1,115,040 sq. mis.), shared by nine countries. Egypt, Sudan, and Ethiopia take up 85% of the basin area while upstream Uganda, Tanzania, and Kenya account for 14%. Three other countries, Rwanda, Burundi, and Zaire share less that one per cent of the basin.

Historian Alan Morehead has told an eloquent tale of the respective discoveries of the sources of the two Niles in the last century. "In these deserts," wrote Morehead, "the river was life itself. Had it failed to flow, even for one season, then all Egypt perished." The fact that it had never failed was one mystery. Another was that it "was possible for the river to continue in its lower reaches for well over a thousand miles through one of the most frightful of all deserts without receiving a single tributary and hardly a drop of rain" (Morehead, 1960, p. 7)

3.1.a. Land Area, Independence Dates of Countries in the Nile River Basin

Basin size (000 sq km)	Length (km)	Country	% of basin land area	Independent Country/Date
3.031	6,632	Egypt	62.7	GB/1956
		Ethiopia	9.9	GB/1937
		Uganda	12.1	NA
		Tanzania	7.7	GB/1961
		Kenya	3.8	GB/1964
		Rwanda	1.8	Belgium/1962
		Burundi		Belgium/1962

During Egyptian control other Nile countries have experienced decades of famine brought on by drought as well as bloody internal conflicts. Increasingly severe water shortages are projected for the future. With most countries already below the annual minimum survival level of 1,000 cms per capita, severe shortages are anticipated in the first quarter of the next century. Despite its river monopoly, Egypt itself will experience a drop in per capita level to 620 cms during this period.

Table 3.2. Economic, Social Indicators, the Nile Basin

	Egypt	Ethiopia	Sudan	Tanzania	Kenya	Uganda
Social Indicators						
Population, millions	62.9	55	28.1	29.7	28.3	21.3
% urban	44.8	13.4	24.6	NA	27.7	29.1
growth, 1978-1995, %	49.6	53.6	70	72.3	104.6	67
infant mortality per 1000 live births	52	120	78	84	59	122
Life expectancy (M)	58	46	52	51	49	44
(F)	60	49	54	55	54	46
Literacy rate, 1995 UNESCO estimate	51.4	35.5	46.1	67.8	78.1	61.8
Economic Indicators						
GDP in $ billions	42.9	5.4	12.1	3.6	8	5.8
GDP per capita in $	660	100	450	140	250	190
Labor distribution						
Industry %	20.5	3	8	5	20	5
Agriculture %	42.7	90	64	82	19	83
Other (services)	36	7	23	13	48	12
Trade $millions						
exports	3,474	202	323	435	1,377	179
imports	9,592	5,293	1,081	1,362	1,696	380

3.2.1 EFFECT OF RIVER BASIN USE

Egypt

Intensive irrigation use of the Nile by Egypt began in the early 20th century. Channels were dug, barrages constructed and the river diverted for larger reservoir storage. Over

96% of Egypt's population of 56 million lives along the narrow, fertile river valley and its delta, only 4% of the country's total land area. Despite its control of river resources, Egypt continues to suffer severe economic and political problems. The nation remains heavily dependent upon foreign aid. The Egypt-Israeli peace summit sponsored by United States President Jimmy Carter in September 1978 has guaranteed annual payments to each country of about $2.5 billion a year.

Aswan Dam

The Aswan High Dam, completed in 1970 with Soviet support after Egypt's falling out with the west and its nationalization of the Suez Canal, has been of some help in increasing the amount of land under cultivation and has provided electricity in significant amounts. Of extraordinary size, its reservoir, one-third of which lies across the border in Sudan, is an average width of 9.6 km (6 miles) and extends upstream for 499 km (310 mis). The Aswan Dam has been the subject of controversy. Its operation has contributed to the deterioration of the Nile delta which is "being eaten by the oceans and invaded by the desert (Shalash, 1983, p. 37 ff; also Burelli, 1992, p.9, and Pearce, 1993, p. 52)." The silt the dam has trapped in Lake Nasser leaves Egypt using more fertilizer per hectare for crop production than is used in any other country in the world. Before Aswan, the receding flood flushed salts away but now they are left behind, and one ton of salt accumulates per hectare each year.

Sudan

Potential competition for Nile water comes primarily from Sudan. Located immediately south and upstream, Sudan is the largest country in Africa, and one of the poorest, with an annual per capita income of $450. Long under joint British and Egyptian rule, Sudan became independent in 1956.

Severe internal problems have resulted from conflicts between the black provinces of the south and the Arab region of the north. During the 1980s civil war and famine caused over half a million civilian deaths. Development funds had also been diverted by the government to the war effort. The major Jonglei Canal project, which was intended to divert the waters of the White Nile away from the swampy Sudd regions of southern Sudan and increase the river's flow into both northern Sudan and Egypt, was abandoned in 1984. Opposition from the southern region contributed to the end of the water project.

Ethiopia

Ethiopia, to the east of Sudan, experienced major famines throughout the 1970s and 1980s, which resulted in the deaths of an estimated two million people. The socialist government tried to alleviate famine conditions in the north by forcing resettlement of more than 500,000 people in the southern region of the country. Application of the Soviet model of agricultural development had disastrous consequences. These policies, which were criticized internationally, were not successful, leading to a sharp decline in agricultural production.

While extractions of river water by Ethiopia would have serious consequences for Egypt and Sudan, internal problems have forestalled such initiatives (Whittington and McClelland, 1992, p. 144-145). Ironically, severe crop losses occur even though more than 80% of the water of the Nile River comes from Ethiopia's highlands each year, as does virtually all of the rich silt reaching Egypt's Nile delta. The famines had a number of related causes. Over-use of the fertile areas in the northern highlands led to deforestation and soil erosion. Civil war in Ethiopia and in neighboring Sudan and Somalia, unexpected climate changes, and an exhausted agricultural base combined to cause the widespread periodic famines.

3.2.2. MAJOR RESOURCE CONFLICTS AND THEIR RESOLUTION

Sudan began at the turn of the century to modify the flow of the river to accommodate changes in agricultural use. Conflicts between Sudan and Egypt over this decision continued despite an agreement consummated in 1929 while both countries were protectorates of the United Kingdom. Sudan was accused of failing to live up to the agreement by raising the level of the Sennar reservoir for its own use. As the disagreement escalated in the late 1950s, Egyptian troops were sent to the Sudanese border. A coup in Sudan in 1958 installed a government favorable to a second Nile agreement for allocation of Nile waters to the two countries.

The 1959 Agreement

The 1959 "Agreement for the Full Utilization of the Nile Waters" allocated all Nile waters to Egypt and Sudan, establishing a Joint Technical Committee to monitor planned projects and conduct hydrologic surveys in the upper Nile region. In anticipation of future problems with the seven other countries in the basin, the two members of the Committee agreed to help to resolve them and to have a common position with respect to them. One of its provisions compensated Sudan for the relocation of 50,000 Sudanese displaced by the construction of the Aswan Dam (Murphy and Sabadell, 1986, p. 140).

Upstream Countries

Many experts agree with Boutros-Ghali that there may indeed be problems with the upstream countries in the years ahead despite the fact that none has yet expressed a need for more water under the terms of the 1959 agreement. Increasing need for water for most of the countries has been documented. One problem is the lack of technical knowledge about the hydrology of the river. "Given the importance of the Nile to Egypt, Sudan, and Ethiopia," it may seem odd that there are technical questions about Nile hydrology and ecology and about development possibilities that have not already been thoroughly studied (Whittington and McClelland, p. 145)." The writers blame this lack on political upheavals and famines in Sudan and Ethiopia and the reluctance of the major donors to become involved in a long-term planning effort when economic and financial conditions are unstable and assistance is required for more immediate problems.

Development Plans
Three possible plans for improving basin-wide access to Nile resources have been suggested, according to Whittington and McClelland. The "Blue Nile Schemes" proposed by the United States Bureau of Reclamation would develop 33 irrigation and hydroelectric power projects in the Blue Nile basin. A second would rehabilitate the pump irrigation schemes on the White Nile south of the Jebel Aulia Dam and draw water from the White Nile throughout the year. A third environmentally oriented proposal would support an exchange program whereby groups in industrialized countries would pay for natural resources preservation measures undertaken by basin countries. None of these development plans are being considered seriously by international funding sources at present.

3.2.4. EVALUATING BASINWIDE AGREEMENTS

Agreement on a forum for discussion and information sharing
The 1959 Nile Waters Agreement remains in effect as the legal basis for the allocation of Nile waters between Sudan and Egypt.
A common forum for the solution of basin-wide issues and the sharing of data has been achieved to a very limited degree and remains at the discretion of the downstream countries. Its provision for responding to the claims of other countries has not been used.
The Technical Co-operation Committee for the Promotion of the Development and Environmental Protection of the Nile Basin, a regional intergovernmental body, has extended its activities to training and water quality monitoring for experts from all the Nile countries.

Binding Convention for all countries
Such a long-term agreement is not yet contemplated.

A basin-wide economic development plan
There is no such plan, despite the urgent need basin experts see for it. Development continues to be Egypt-dominated.

Sharing of data
Hydrologic monitoring is conducted by Sudan and Egypt as part of their 1959 agreement. The Foreign Broadcast Information Services reported Sudan's refusal to admit two high level Egyptian government representatives into Sudan in June 1994.
Sharing of compatibly developed water quality or other environmental measurement data basin-wide is lacking.

Development and donor sensitivity to environmental impacts
Because of the lack of investment activity there have been no recent pportunities to test donor attitudes.

3.3. The Senegal: Basin Countries Hold Firm

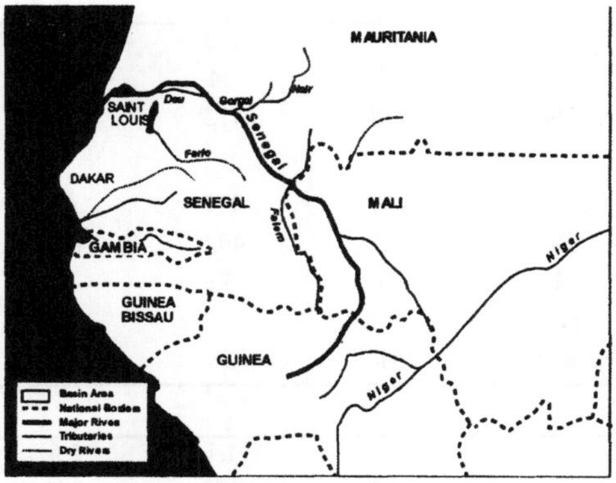

The Senegal River Basin

The four countries in the Senegal River Basin have supported basin-wide management since the 1960s, challenging donor development plans and seeking to resolve water use issues in other ways. Their cooperation has had mixed results, reflected by continued poverty in the region, armed conflicts, and failure to qualify for international funding.

3.1.b.Land Area, Independence Dates of Countries in the Senegal River Basin

Basin size (000 sq km)	Length (km)	Country	% of basin land area	Independent Country/Date
0.353	1,635			
		Mali	46.2	France/1960
		Mauritania	26.3	France/1960
		Senegal	19.1	France/1960
		Guinea	9.3	France/1958

3.3.1. BASIN PROFILE

The river is 1,600 kms. long (1,000 mis.), with a drainage area one-tenth the size of the Nile Basin. It rises in Guinea and includes in its basin three other former French West Africa countries, Mali, Senegal and Mauritania. Like the Nile, much of the basin lies within the Sahel, the semi-arid region which stretches across Africa from Senegal to Ethiopia, and is thus affected by severe drought. All four states became independent of France in the 1960s. They suffer severe poverty, primarily the result of a climate which dooms most of their inhabitants to subsistence-level farming. Crops for export, tourism, and some industry continue at a low level.

Mali

Mali is sparsely settled, with most of its population clustered along the Niger and Senegal rivers. Continuous periods of drought have plagued farmers. Subsistence agriculture supports the economy, although only about 2% of the land is suitable for

Table 3.3 Economic, Social Indicators, the Senegal

	Mali	Mauritania	Senegal	Guinea
Social Indicator				
Population, millions	10.9	2.3	8.3	6.7
% urban	27	54	42.3	29.6
growth, 1978-1995, %	59.1	47.2	53.1	28.9
Infant mortality				
per 1000 live births	159	117	68	134
Life expectancy (M)	55.2	49.9	48.3	44
(F)	59	53	50	45
Literacy rate, % +15 pop.				
1995 UNESCO estimate	31	38	33	36
Economic Indicators				
GDP in $ billions	2	1.1	5.4	2.3
GDP per capita in $	250	480	750	520
Labor distribution				
Industry %	NA	NA	NA	NA
Agricultural %	82	66	77	78
Other (services)	18	34	23	22
Trade $millions				
exports	301	507	661	590
imports	666	581	1,384	775

cultivation. The drought that began in 1983 drastically reduced crop output as well as the availability of fish, a staple food, in the Niger, forcing nomads deprived of large herds of cattle, sheep, and goats into refugee camps. Mali shares a chronic trade deficit with its neighbors, along with a heavy dependence on foreign aid.

Senegal

Senegal, the westernmost country in Africa, has a 600-km. (373-mi.) Atlantic coastline and is bordered on the south by Guinea-Bissau and Guinea, on the east by Mali, and on the north, across the Senegal River, by Mauritania. Its capital, Dakar, is one of West Africa's leading ports. Compared with most of its African neighbors, the country has a relatively well-developed industrial sector, with manufacturing accounting for about 20% of the GNP. Attention has focused chiefly on the processing of local raw materials and the production of consumer goods, textiles, chemicals, and petroleum products. Tourism provides a significant source of foreign exchange. Senegal usually runs a large trade deficit, however, and economic problems have contributed to political unrest.

Mauritania

Mauritania has also suffered from severe drought. The result has been a massive migration by its cattle-herding nomads, who constituted 83% of the total population in 1963, and had been reduced to only 25% by 1986. Eighty per cent of the population and most of the nomads are Moors, descended from Arabs and Berbers. Before the

Sahel drought of 1969-74, Mauritania's farmers grew enough grain to feed the country. Since then, however, most foodstuffs have had to be imported. Due to drought and overgrazing, the desert is moving southward at a rate of about 6.5 km. (4 mis.) per year. Efforts to restore pastureland by digging wells and stabilizing dunes have met with little success. The country is heavily dependent on food imports and foreign aid.

In 1989 a border incident with Senegal resulted in large-scale ethnic violence and the displacement of thousands from each country, including the expulsion of many black Mauritanians to Senegal. Most recently, friendly relations between the two countries have been restored.

3.3.2. MANAGING RESOURCES AND CONFLICTS

Improper use of Senegal River natural resources among other factors has doomed its basin population to poverty levels that are among the lowest in Africa. Improved river development is key, in the view of resident experts. An authority on the Senegal, David G. LeMarquand, sees improved river management as the way to avoid the "curse of climatic and hydrologic unpredictability." Regulation of river flow could reduce the waste of resources in annual flooding and the long months of low flow." A regulated flow can provide water for year round irrigation to help make up the deficit in rice and other staple crops; it can maintain depth for year round navigation up to 800 km. inland; and it can generate hydro-electric power for countries suffering greatly from dependence on imported oil (LeMarquand, 1987, p. 8)."

The three countries have agreed to preserve freedom of navigation, a right limited to the contracting states. National projects must obtain the approval of OMVS if they would affect the river's rate of flow, navigability, or endanger public health by reducing water quality or the use of water for agriculture and industry. The OMVS has a three-level structure, including a governing body of heads of states, a council of ministers to elaborate general policy, and a secretariat, to implement decisions. More recently a High Commission was created to represent the Council of Ministers in its relations with international aid agencies, as well as a committee to harmonize national agricultural research and development programs and a committee to advise on financial matters (OMVS Magazine, 1995). A permanent commission on water had been added in 1976.

While countries await funding to complete the agreed upon development plan other ethods for improving river resource use have been put in place. Farmers in the Senegal River valley practice flood-recession agriculture, an indigenous cultivation method used by millions that involves planting crops on river floodplains after seasonal floodwaters recede. "By engineering controlled floods," according to resource specialist Sandra Postel, "the reliability and productivity of this method can be improved without ecological damage, exorbitant cost and disruption of local traditions that full-scale irrigation often brings (Postel, in Gleick, 1993, p. 63)."

3.3.3. EVALUATING BASINWIDE AGREEMENTS

Agreement on a forum for discussion and information sharing
The four countries in the basin agreed on a forum approach to basin management soon after gaining independence from France.
While Guinea has not participated in agreements and discussions on occasions, a basin-wide pact has supported the creation and operation of decision making and technical units with specific tasks and functions.
The primary institution, the OMVS, survived political disputes among the three countries and disagreements over some of its main functions.

Binding Convention for all countries
See above. The OMVS is a potential instrument for increased and continued cooperation on all aspects of river basin management.

Basin-wide development plan
A basin-wide plan has been the focus of the Senegal basin countries and subjected to scrutiny by experts and politicians in each country. It has been put in place to a limited degree and awaits further funding.

Sharing of data
This takes place at the technical level but the degree of technical competence and compatibility is uncertain.

Development and donor sensitivity to environmental impacts
The OMVS has contributed to an increase in sensitivity through its insistence that measures protecting agricultural and fish resources be maintained.

3.4. The Lower Mekong: Multiple Conflicts, a Final Development Plan

Like the Senegal, cooperation in the Mekong River Basin has focused on river development plans. Planning activities have been concentrated in the lower Mekong, and have not included the two upstream countries, China and Myanmar (Burma). Cooperation among the countries in the lower basin, Cambodia, Laos, Thailand, and Vietnam, including agreements on navigation and other basin uses, has survived a series of bloody ideological conflicts. Like the Senegal, the Lower Mekong needs development which will help to regulate irrigation, particularly for staple rice crops.

The Mekong rises in China, in eastern Tibet, and flows south from there, draining an area of 785,500 sq. km. The eleventh longest river in the world, it crosses or forms the boundary of five countries south of China, Myanmar (Burma), Laos, Thailand, Cambodia and Vietnam, covering a distance of 4,500 kms. (2,750 mis.). It empties into the South China Sea from Vietnam via a vast, complex and fertile delta of 194,250 sq. kms. (75,000 sq. mis.).

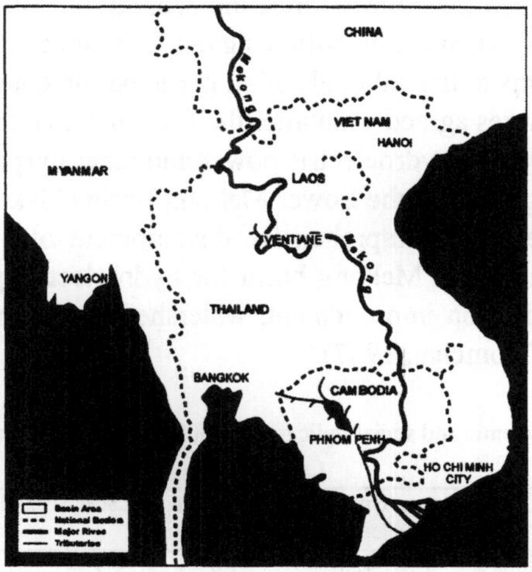

The Lower Mekong Basin

3.1.c. Land Area, Independence Dates of Countries in the Mekong River Basin

Basin size (000 sq km)	Length (km)	Country	% of basin land area	Independent Country/Date
0.786	4,180	China	22.2	NA
		Cambodia	19	France/1953
		Laos	25.4	France/1953
		Vietnam	7.7	France/1954
		Myanmar	2.9	GB/1937
		Thailand	22.9	NA

Over 50 million people use the lower river and its tributaries for transportation, food, drinking water and irrigation. Mekong floods each year have provided support for the rice-based agriculture of riparian countries for millennia. These floodwaters also nourish wetlands and bottomland forests, providing a food chain for a fish harvest of approximately 700,000 tons each year (Rothert, 1995, p. 5). The eastern basin region (Laos, Cambodia, and Vietnam) became known as French Indo-China in the second half of the 19th century. Elsewhere in the basin, Myanmar was a British colony until independence in 1948 and Thailand, which had been an independent monarchy since the 14th century, became a constitutional monarchy in 1932.

3.4.1. DEVELOPMENT BY AGREEMENT, 1957 - 1995

Despite the history of armed conflict within and among the lower Mekong River Basin countries, potential disagreements over some basin uses have been resolved through international agreements for several decades. Freedom of navigation on the lower Mekong, which had been guaranteed since the second half of the nineteenth century, was reaffirmed in 1950 when a convention signed by France, Cambodia, Vietnam, and Laos, granted such rights to the nationals of all participating countries.

The four lower countries agreed soon after the French defeat to join in plans for the development of the river for hydroelectric power and other purposes. They formed the Committee for Investigations of the Lower Mekong Basin (Mekong Committee) in 1957. Its goal has been the "comprehensive development of the water resources and related resources of the lower Mekong basin for hydroelectric power, irrigation, flood control, drainage, navigation improvement, watershed management, water supply and related development (Chomchai, 1987)."

Table 3. Economic and social indicators for international river basin countries, Lower Mekong

INDICATOR		Burma	Cambodia	Laos	Thailand	Vietnam
Social Indicators						
Population, millions		42.6	7.3	4.4	57.7	69
% urban		24	13	16	18	19
% rural		76	87	84	82	81
annual growth, %		1.9	2.2	2.9	1.4	2
Infant mortality, per 1000 live births		68	121	107	35	47
Life expectancy	(M)	57	48	49	67	63
	(F)	61	51	52	71	67
Literacy rate % adult pop		81	35	41	93	88
Economic Indicators						
GDP in $ billions		22.2	0.93	0.8	92.6	15
GDP per capita in $		530	130	200	1,630	220
Labor distribution						
Industry %		11		2		14
Agricultural %		65	70	80	58	73
Other (services)		20	30	18	22	5
Trade $millions						
exports		568	17.4	72	27.5	1,800
imports		389	513	1	692	1,900

The Committee, supported by the UNDP and other sources, prepared, on the basis of hydrologic and meteorologic data collected over a period of several years, a "Lower Mekong Water Resources Inventory" to support a basin plan for the final three decades of the century. The plan proposed the construction of 116 possible mainstream and 180 possible tributary, projects. A secretariat was created for the Committee in 1963 with funding from the UNDP.

In 1961 a Japanese consultant completed a reconnaissance of the lower Mekong which, according to Hiroshi Hori, a Japanese expert writing in International Water, "was regarded as a fundamental guidebook for all following studies carried out by the Mekong Secretariat and its cooperating countries (Hori, 1993, pp.110-115)." The study suggested the construction of ten dams built as cascades to control flooding, generate hydropower of about 19-20 million KW, provide irrigation for more than a million hectares, and improve navigation.

Committee Activities

The Committee has undertaken a number of management activities, recruiting staff from the member countries. An extensive network of hydrologic and meteorologic stations has been established for operations and planning purposes. Water quality and water balance monitoring systems also have been put in place and training programs established for technicians and farmers. Development funds for the region of $600 million were provided by various donors. Twenty-five per cent has been used for pre-investment activities and 75% for investment in construction.

Development plans got a major boost in April 5, 1995, when the four original Mekong Committee members signed the "Agreement on the Cooperation for the Sustainable Development of the Mekong River Basin" which formally converted the Mekong Committee into the Mekong Commission, and reactivated the Committee's regional planning authority. The International Rivers Network has provided a critical analysis of the plan supported by the new commission, in particular what its author sees as its unfavorable impact on the basin's natural resources (Rothert, 1995).

The newly named Commission's basin-wide activities have resulted in an extensive network of hydrologic and meteorological stations (392 hydrologic and 332 meteorological), spanning a large part of the lower Mekong basin, to collect basic field data needed for essential operations and planning purposes. Changes in basin-wide water quality and water balance are being assessed as monitoring systems are established. By 1996 little water project construction had taken place in the lower Mekong despite decades of planning.

3.4.2. EVALUATING LOWER MEKONG AGREEMENTS

Agreement on common forum to solve problems and share data

The common forum for agreement in the Mekong River Basin has been confined to the lower Mekong, excluding Myanmar and China.

The four countries in the lower basin have negotiated agreements on navigation and other matters since the defeat of France and the break-up of French Indo-China in 1954.

Binding Convention for all countries

The agreement signed by the four countries of the lower Mekong in 1957 created a Committee to implement a number of goals related to the management of the lower basin.

The Committee has had a secretariat and other policy and operating units since 1963.

The Committee, now Commission, reaffirmed its role by agreement on a number of development plans including the most recent in 1995. Agreement will no doubt be confined to the lower Mekong for the foreseeable future.

Basin-wide development plan

The lower basin has had a succession of development plans from various sources.

The most recent proposal (April 1995) relies on more sophisticated data than has been available in the past but is accused by environmentalists of not responding to the need for detailed environmental impact assessments.

Development and sharing of data

The Lower Mekong Commission has increased the number of hydrologic and water quality monitoring stations but it appears that the latter exist in small numbers; information has been lacking about the amount of data, particularly water quality, collected and their availability.

Development and donor sensitivity to environmental impacts

The UNDP has provided advice and financial support for the Commission which, according to the World Rivers Review of August 1995 "has made no provision for public participation or observation of its activities, nor have minutes of its proceedings been made publicly available."

With more than 100 major dams proposed for the basin, critics view the Commission, and UNDP's role in it, as a "front for dam construction proponents (Lammers, 1995, p.4)." Donor sensitivity from this quarter at least is in question.

3.5. The De la Plata: Development by Agreement?

The De la Plata, the fourth largest river basin in the world, drains an area of 3,200,000 sq. kms. (1,115,040 sq mis.). Its two major river systems, the Paraguay/Paraná and the Uruguay, empty into the Atlantic Ocean via the De la Plata River. The Basin supplies extensive resources to Argentina and Brazil, the two largest and most developed countries of South America. The complex network of basin rivers provides transportation for the landlocked countries of Bolivia and Paraguay and hydroelectric power for Argentina and Brazil and to some extent the other three countries. The divergent needs of the basin countries have caused serious conflicts which have focused on development issues.

3.5.1. BASIN PROFILE

The Paraguay/Paraná system is a major resource for Paraguay, Brazil, Argentina, and Bolivia. It drains an area of 1.75 million sq. kms. The Parana River, 4,880 sq km (3,030 sq miles), is South America's second-longest river after the Amazon. It begins in south-central Brazil and drains an area of approximately 2.8 million sq km (1,080,000 sq mis.). The Paraguay River is 2,550 kms. (1,585 mis.) long, rising in south-central Brazil, and flowing south into Paraguay where it forms the border between Paraguay and

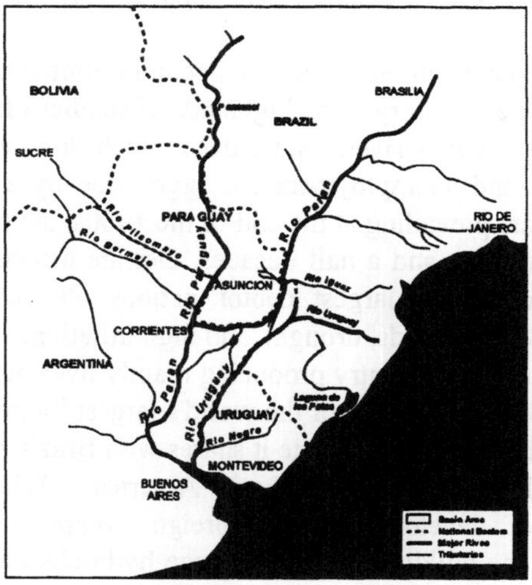

The De la Plata Basin

Table 3.1.d. Country Basin Share, Independence Dates, De la Plata (Parana)

Basin size (000 sq km)	Length (km)	Country	% of basin land area	Independent Country/Date
3.2	4,880	Argentina	30.9	Spain/1816
		Brazil	44.5	Portugal/1822
		Paraguay	12.7	Spain/1811
		Bolivia	7.4	Spain/1825
		Uruguay	4.4	Spain/1828

Argentina. It is a tributary of the Parana, joining it north of Corrientes, Argentina. The Paraguay drains a relatively empty and little-developed region of 1,100,000 sq km (425,000 sq mi). Of major importance to landlocked Paraguay, it is navigable by large vessels as far as Corumba, Brazil, on the Bolivian border.

The Uruguay river is 1,610 kms. (1,000 mis.) long, draining over 300,000 sq. kms. (118,530 sq. mis.). It rises in southeastern Brazil and flows west and southwest, first along the Argentina-Brazil border and farther south along the Argentina-Uruguay border before joining the de la Plata River. It is navigable for about 200 mis. upstream and serves Uruguay, Paraguay, and Brazil .

Four of the five basin countries were Spanish colonies before achieving independence in the early 19th century; Brazil was a Portuguese colony until the same period of time. While each has experienced political instability, their economic and social indicators, with the exception of Bolivia, are substantially higher than countries in the Nile, Senegal, or Mekong (excepting Thailand) basins.

Argentina is the second largest country in Latin America and one of the more highly developed countries in the Western Hemisphere. It occupies most of the southern portion of the South American continent and is the third most populous country in Latin

America (after Brazil and Mexico), containing about 7% of that region's total population.

Brazil, the largest country in the continent, occupies almost half of its land mass and is the world's fifth-largest country. The highlands of southern Brazil are drained by the Paraná, Paraguay and Uruguay rivers, water from which flows eventually into the De la Plata. Urban poverty and unemployment are aggravated by a fast population-growth rate. At the growth rate prevailing at the end of the 1980s, the population of Brazil was expected to double in three and a half decades. Despite a potentially strong economy Brazil is one of the world's largest debtor nations; its economic problems were aggravated in the 1980s by floods, drought, and high inflation.

Paraguay is an agricultural country producing mainly livestock and cotton. In the late 1970s and early 1980s, construction of the world's largest hydroelectric installation, the giant Itaipu Dam on the Paraná River site it shares with Brazil, helped to give Paraguay the highest economic growth rate in Latin America. When construction ended, however, the rate of development and foreign investment in Paraguay slowed dramatically, despite the building of another huge hydroelectric project on the Paraná, Yaycreta Dam, which was expected to become operational in the mid-1990s. Factories built in the 1970s were operating far below capacity, and unemployment increased steadily in the 1980s.

Table 3.5 Economic, Social Indicators, the De la Plata

INDICATOR		Argentina	Brazil	Bolivia	Paraguay	Uruguay
Social Indicator						
Population, millions		34.6	161.8	7.4	5	3.1
% urban		88.1	78.3	60.8	52.7	90.3
growth, 1978-1995, %		26.7	38.7	40.1	63.7	10.8
Infant mortality, per 1000 live births		23	56	75	38	19
Life expectancy	(M)	67	64	57	64.8	69
	(F)	74	70	61	69.1	76
Literacy rate, % + 15 pop 1995 UNESCO estimate		96	83	83	92	97
Economic Indicators						
GNP in $ billions		306.1	600	7	8.9	15.8
GDP per capita in $		8,110	2,970	770	580	4,660
Labor distribution						
Industry %		30.5	25	14	34	31
Agricultural %		12	35	50	44	16
Other (services)		57.5	40	26	22	22
Trade $millions						
exports		15,800	43,355	1,005	817	1,918
imports		21,581	35,548	1,182	2,425	2,707

In Uruguay increased urbanization has intensified agricultural development; the country is largely self-sufficient in food production. In Bolivia malnutrition is widespread and cultivable land is scarce; much of the better land is terraced, and the terraces, like the farming methods, date from pre-Inca times. Herds of sheep and llamas are raised on the drier, southern parts of the Altiplano, and irrigated agriculture is practiced in the south where fresh water is available.

3.5.2. DEVELOPMENT: 1960S TO THE 1990S

The Brasilia Treaty of 1968 signed by all five countries set up a mechanism to assist basin economic development. It established the Intergovernmental Coordinating Committee of the Del Plata Basin Countries (CIC) to implement the agreement (Cano, 1987, p. 7). The CIC meets frequently and has a permanent secretariat. There is in addition a financial fund to support basin development (FONPLATA) which was established in 1976. It is capitalized by the countries themselves with help from the United Nations Development and Environment Programmes, the Office of American States and other international agencies. For the most part conflicts over development have been resolved bilaterally.

The Itaipu Dam

The Paraná has been of special interest to Brazil since the mid-1950s when the network of small hydroelectric projects on nearby rivers and on the Parana itself failed to supply enough energy for the country's continued development (Murphy and Sabadell, 1986, p. 133). Brazil's need to overcome fossil fuel shortages and to ease its dependence on imported oil placed a special high priority on the construction of a project at the Itaipu site, near the gigantic Falls of Guaira.

Argentina objected to the Itaipu project because it would have interfered with its own plan to join Paraguay in water projects on their shared reach of the river. Brazil and Argentina eventually reached an agreement to coordinate their projects. Their lengthy negotiations over the dam included a show of force by Brazil which had initially claimed exclusive sovereignty over the site. Both countries signed the Itaipu Treaty in 1973 agreeing to build Itaipu. Now operating, Itaipu has a capacity of 12,600 MW, more than five times the capacity of the Aswan Dam. The 8 km (5 mile) long dam provides most of Paraguay's electricity and power for industrialized southern Brazil. Construction costs exceeded $20 billion.

An administrative council and an executive directorship were established to implement the agreement to build the dam, each composed of an equal number of representatives from each country. The countries agree to buy, together or separately, the total amount of energy produced, eliminating the possibility of exporting energy to third parties. Paraguay has had to pay half the cost of construction and was disadvantaged by substantial cost overruns. It exports a substantial share of the dam's electricity to Brazil. A number of Brazilians have migrated to Paraguay territory around the dam: farmers, engineers, technicians as well as the construction workers. The total is several hundred thousand, more than one-tenth of Paraguay's population.

A second major hydropower project, the Yacyreta on the Paraguay River, has been guided by an agreement between Argentina and Paraguay which created a special binational agency to develop funding and supervise construction. The project has been entirely financed by Argentina with benefits equally divided between the countries.

The Paraguay-Paraná Hidrovia Project

River basin development faltered in the 1980s due to a lack of funding and poor economic conditions. Some river projects' supporters believe that the signing of a treaty by Argentina, Brazil, Paraguay and Uruguay to establish a common market for their countries might encourage the development of water projects now on drawing boards. The agreement, called Mercado Comun del Cono Sur or Mercosur, was signed in March, 1991. The common market covers a region of 12 million sq km with a population of almost 200 million people and has a potential capacity of producing goods and services believed to be worth $630 billion a year (Lammers et al, 1994, 27). Institutional arrangements have included the establishment of the Common Market Group which administers the program. An administrative secretariat is located in Montevideo, Uruguay. Mercosur has received funding from the Inter-American Development Bank (IDB) of about $4 million.

Disagreements delaying the implementation of the plan have focused on the Hidrovia (Waterway) Project which would improve transportation on the Paraguay-Paraná river system by making improvements to existing port, channel and navigation facilities and the construction of a year-round navigable waterway from the headwaters of the Paraguay River (in Brazil) to the Uruguayan port of Nueva Palmira in the river's delta on the Atlantic. The waterway would be over 3,400 kms. (2,100 mis.) long. Its planning has been funded primarily by the IDB and the UNDP, contributing toward an estimated start-up fund of $11 million.

A number of basin-wide and international nongovernmental organizations have questioned the high financial and environmental costs of the waterway. Alternatives, a series of smaller projects, are under consideration. The possible danger to the Pantanal, the world's largest wetland, through which the Paraguay flows and most of which is located in Brazil, is causing wide concern among conservationists. A number of threats to the Pantanal from mining waste as well as deforestation in Paraguay have also been noted. A vigorous campaign is under way to gain to achieve full public participation in the planning processes related to the waterway (International Rivers Network, 1995).

Environmentalists point out that despite the large numbers of bi- and multi-lateral agreements and commissions which have been established over the years to address the management of transboundary ecosystems and environmental problems within the De la Plata River Basin, "none have been particularly effective, given the decline in water quality in the rivers and the continued degradation of the Pantanal wetlands (Lammers et al, 1994, 31)."

3.5.3. EVALUATING DE LA PLATA RIVER BASIN MANAGEMENT

Agreement on common forum to solve problems and share data:

A common forum for discussion purposes has brought the De la Plata countries together from time to time; available data have been shared during those discussions.

Economic rivalries and the perceived need for large-scale navigation and hydroelectric projects have resulted in bi- and tri-lateral agreements.

Binding Convention for all countries

So far, serious conflicts over the development of Basin resources have been settled bilaterally.

There have been two important basin-wide agreements, one establishing the CIC (Coordinating Committee of the Del Plata Basin Countries) to foster joint development and one to establish a common market in a substantial part of the basin.

Political and economic difficulties are obstacles to continuous basin-wide management. Recent widespread attention to environmental concerns have encouraged discussion but effective management agreement has not been reached.

Basin-wide development plan

The size and complexity of the river system makes implementation of a unified development plan extremely difficult.

On the other hand the proposed development of the Parana-Paraguay water way project encompasses a major part of the basin. It and the Mercosur may lead to closer cooperation of the countries in all phases of development, including water resources planning and use.

Development and sharing of data

For the most part data-sharing is limited to exchanges about individual projects and navigation.

Development and donor sensitivity to environmental impacts

The possible threat to the El Pantanal wetland has stimulated interest among both donors and national governments.

The key ingredient, application of objective environmental impact assessments which include wide public participation, is being supported but is far from implementation.

3.6. Summing Up

Stephen C. McCaffrey (McCaffrey, 1993, p.103, endnote #105) reports that there are fourteen multilateral agreements concerning international water courses, both lakes and river basins. All but one were concluded after 1950. His list includes the Danube, the Rhine, the De la Plata, and the Senegal. He also includes the Mosel, Niger, Kagera, Gambia, Amazon and Zambezi river systems. The Nile and the Mekong presumably have been excluded because agreements cover only a part of the basins. The sample of river basins used in this chapter makes it possible to contrast accomplishments with needs in a significant number of international river basins which have multi-lateral agreements.

The need for agreement on equitable sharing of river basin resources is obvious in the four basins in developing regions. Their inability to satisfy the needs of increasing populations is complicated by factors beyond the control of an agreement on sharing resources. Agreement, in turn, depends on the resolution of difficult historic, economic, political, and cultural imbalances beyond the scope of a basinwide management convention.

Conclusions based on their respective agreement achievements

Negotiation forum: Failure to engage in discussions, with or without a convention, is a significant factor in basin management. With formal or informal discussions supported by the countries themselves the opportunities for advancement to a second level of agreement are enhanced. Leadership by an individual head of state, or by a third-party agency can be important at any time in the agreement process.

SCOPE OF EXISTING AGREEMENTS

Formal conventions which go beyond navigation and extend to control of pollution or other harm to the river from the activities of some countries at the expense of others are rare. In our sample of four all have formal conventions on the use of the river's water. These are limited in substance in each case, however, and do not extend to the kind of agreements in force in the Rhine Basin, and now anticipated for the Danube. They are also all involved in the use of the river for economic development. Incorporated into some of the agreements, however, are provisions for taking environmental effects into consideration, notably in the Senegal Basin.

Sharing data:

Key to management of water resources is a network of data monitoring and sharing. A management plan is useless without accurate, compatible measurements of quantity and quality levels. Some basins have provided for limited data-sharing. The possibilities for applying state-of-the-art technology for both monitoring and shared reporting need to be examined and their development funded. Training, which is already available in a limited way, for developing and using the technology would be an urgent follow-up.

Basin-wide plans:
Planning for the most part has come from the need to develop rather than manage the basin for equitable use by all basin countries. In the Nile basin the two significant parties to agreement, Egypt and Sudan, planned and carried out a monopoly of basin resources with minimal attention to other counties. Third parties seeking to develop parts of the basin have been a major factor in plans which for the most part concentrate on large hydro-electric power projects. Countries can protest and cause a renegotiation but this happened in only one of the four basins. Third-party sensitivity to the effect of projects on wetlands and other ecological resources is increasing. The Global Environment Facility begun in 1991 by the World Bank, and the UN Development and Environment Programs is an example. Many environmental groups continue to criticize the lack of effective environmental impact assessments, with public participation.

The receptivity of the countries in these basins depends, as it does elsewhere, on political and economic stability. Not one region can be called stable with respect to either factor. Thus, it has been difficult to achieve the required confluence of domestic and foreign policy (Chapter 10). Agreements under such circumstances lack permanence and equity.

CHAPTER 4

A European Model For The Danube? The Rhine

4.1. Basin Profile

The economic and social indicators of Rhine River Basin countries show striking differences from those of the countries of the four basins in developing regions. Age expectancy in each of the four riparian countries is in the upper seventies. Infant mortality rates (per thousand at birth) are low single rather than double and triple digit numbers. Annual per capita income significantly exceeds even that of the most industrialized countries in the less developed river basins. The Rhine Basin countries have reached and implemented agreements on its management, most recently to curb the pollution which severely degraded river resources during the post-World War II industrial boom. Catastrophes, including a major chemical spill and severe flooding, have given notice from time to time that management tools supported by agreements were inadequate.

The Rhine, the longest river in western Europe at 1,320 kms. (825 mis.), rises in the Swiss Alps and flows across Switzerland, western Germany, and the Netherlands to the North Sea. It drains an area of 185,000 sq. kms. (71,429 sq. mis.). For much of its length it forms the boundary between France and Germany. Navigable from Schaffhausen in Switzerland, it passes through large industrial centers from Basel downstream. Despite heavy industrialization in its countries, crop cultivation continues to be extensive. The lower Rhine, from Bonn to the Dutch border, is wide and sluggish. The Rhine has the heaviest water traffic of any river in the world and includes a large number of canals and operational systems projects for the regulation of the water level.

At Remagen the riverbed widens and crosses the plain of northwest Germany. In the early 19th century much work was done on the middle reach of the river so that it could be both straightened and deepened. The river is linked by canals with the Ruhr mining and industrial region. A tributary, the Moselle, connects it with markets in France. The Rhine-Main-Danube Waterway, completed in 1992, now links the Rhine with the Danube River and the Black Sea (Bryson, 1992, p. 8).

4.1.1. POST WORLD WAR II DEVELOPMENT

The Rhine valley played a significant role in the recovery of West Germany after World War II. By 1955 the Republic had become a member of NATO, later participating in the formation of and joining the European Community. By 1970 Germany was one of the world's primary industrial powers, maintaining a high growth rate and a comparatively low rate of inflation. Its vigorous economic recovery relied on the further development of the Rhine for navigation and hydroelectric energy which began in the years between the two world wars.

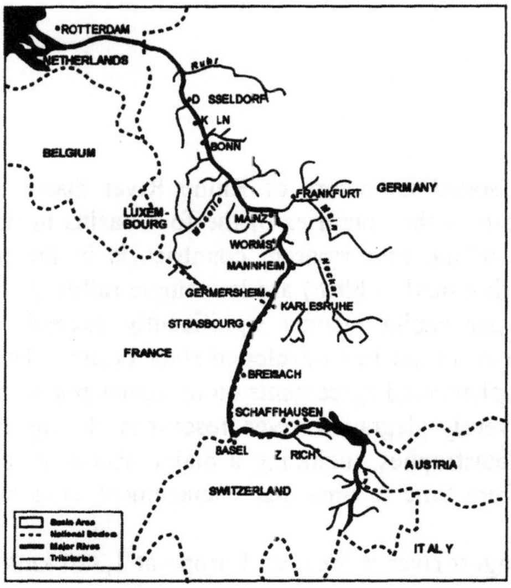

The Rhine River Basin

The Rhine delta begins as the river crosses the Dutch border, dividing into three separate rivers with a number of branches, one of which skirts the port city of Rotterdam, the world's largest port, accommodating the transhipping of over 250 million tons of goods each year (Hupkes, 1990, p. 21). A renowned network of canals and dikes provides an artificial drainage system to keep the land dry and inhabitable. A canal constructed in 1938 links the city of Amsterdam to the Rhine.

4.1.2. RHINE COMMISSION

Congress of Vienna provisions for freedom of navigation on Europe's waterways applied to the Rhine. In 1831 the Central Commission for Rhine Navigation was formed by agreement among all riparian countries to support free transport. A commission to protect fishery resources was established in 1885. Increased concern about pollution in post-World War II stimulated discussion about its impact on the river at meetings of the fishery commission and elsewhere. The International Commission for the Protection of the Rhine against Pollution (ICPR) was formed in 1950 and became an official international organization in 1963. The four riparian members were joined by Luxembourg and, in 1976, by the European Community.

An Open Sewer

Despite the agreement to reduce pollution, as Alexander Kiss, a river basin expert, has commented, there was "the growing consciousness that the romantic Rhine, home of

Table 4.1. Economic, Social Indicators, Rhine River Basin

	France	Germany	Luxembourg	Netherlands	Switzerland
Social Indicators					
Population, millions	58.0	81.6	0.4	15.5	7.0
% urban	72.8	86.6	89.1	89.0	60.8
% growth, 1977-1995	9.2	NA	11.6	11.5	11.0
Infant mortality					
per 1000 live births	4.9	6	5.5	6	5.1
life expectancy, at birth (M)	73.8	73.3	73	74.5	75.3
(F)	79.8	79.8	80	80.2	81.7
Economic Indicators					
GDP, total in $ billions	1,516.1	2,122.1	12.4	385.0	305.1
per capita $	26,127	25,993	30,859	24,992	24,522
Trade					
exports, in $billions	287	508	12.9	196	78
imports, in $billions	275	448	11.8	176	77

Lorelei, was being transformed into an open sewer (Kiss, 1985, p. 614)." Improvement did not come for several years. The downstream Dutch expressed the most serious concern as they measured and reported on extensive quantities of pollutants at the Rhine border point. They found, for example, that the river carried a yearly average from 1973 to 1975 of 47 tons of mercury, 400 tons of arsenic, 130 tons of cadmium, 1,600 tons of lead, 1,500 tons of copper, 1,200 tons of zinc, 2,600 tons of chromium and 12 million tons of chlorides (Hupkes, 1990, p. 24).

Rotterdam Measurements

River pollutants end up in Rotterdam Harbor, in the Rhine estuary on the North Sea. The world's largest port, it must have 23 million cubic meters of harbor sludge dredged each year to accommodate the 250 million tons of goods transhipped through the harbor. About 10 million cubic meters are so polluted they cannot be dumped into the North Sea but must be retained in a specially built site. The Dutch warn that the site will be full by 2002. Ian Dogterom, director of the International Centre of Water Studies in Amsterdam, has observed that the Rhine cleanup which took place during the 1960s was effective to a degree but that by 1988 it was ready to be made clean again (Durkee, 1989, p. 1).

4.2. Dealing with Pollution

The difficulties in cleaning up the Rhine and maintaining its usefulness for drinking water and for other uses have been dealt with in different ways by the Commission, which has its headquarters in Koblenz, Germany. A IIASA report points out that "all five nations realized that among themselves there were both conflict and joint interest, and a common desire to find mutually acceptable solutions. The appropriate format for these negotiations was conference diplomacy--frequent contacts between and formal meetings of representatives (Dupont 1993, p. 145)." The organization has been useful both as a permanent secretariat and advisory body. While some member states, particularly the Netherlands, support institutional change to reinforce implementation of river clean-up, this has been resisted by a majority of members.

4.2.1. COMMISSION STRUCTURE

The commission was established primarily as a forum for collective decision making and to provide information, technical expertise, administrative support, and other guidance to member countries. Decisions are made at annual meetings of ministers of environment or foreign affairs. A multistage process takes place, with the secretariat drawing up proposals, coordinating plans and recommendations, and providing secretarial and technical assistance. Twenty five working groups are made up of experts from each of the countries. Their actions may be followed by formal treaties or some other type of document affirming agreement.

4.2.2. TWO DISASTERS TEST AGREEMENT

Two disasters in the last ten years, pollution from the 1986 Sandoz plant chemical fire and the winter floods of 1994-95, brought the question of Rhine management into sharp focus. The first event produced an action plan and the second has been the occasion for a re-examination of basin-wide flood prevention (Pearce, 1993a).

Sandoz Spill

The fire at the giant Sandoz chemical plant in a suburb of Basel, Switzerland began in two storehouses filled with insecticides, herbicides and fungicides, including highly flammable solvents and dyes, in the early hours of Saturday, November 1, 1986 (Deininger, 1987, p. 81). It took several hours to bring the fire under control and in the process thirty to forty tons of the toxic substances were washed into the Rhine. The Swiss called German and French water authorities but did not know the magnitude and toxicity of the released substances.

On the following Monday a monitoring plan and emergency procedures were developed and by Tuesday a list of the substances stored at the site was made public. Massive fish deaths occurred in the river. Substantial amounts of one chemical, rhodamine B, colored the toxic plume red so that it could be followed down river. The warning systems used by the water utilities along the river were activated. Plume travel-times were estimated and warnings issued so that the authorities could stop taking Rhine water until the plume had passed.

The accident has been described by Linda Durkee of the Bureau of National Affairs as changing "the face of environmental protection in Western Europe." It put "into sharp relief the consciousness about the safety of the highly industrialized corridor along the Rhine." Other recent experiences such as the Chernobyl explosion in Ukraine in 1986 played a part. "But it is from the time of the Basel accident that a far clearer picture of the international implications of environmental protection emerged, and it is from this time that communication of risk about all the pollution practices involving the Rhine took an exponential leap forward (Durkee, 1989, p. 1)."

The countries met immediately after the Sandoz incident to agree on an action plan to substantially reduce pollution in the river by the year 2000. According to commission member Marc Braun, the Rhine Action Plan (RAP) sets goals for the reduction of emissions and fixed target values for about 50 potential pollutants. The target values accommodate acceptable water quality levels for drinking water, fisheries and habitat. The annual monitoring programme samples and tests ninety parameters in water and forty in suspended matter. A more extensive five-year program records the contents of noxious substances in water, suspended matter, sediments and organisms (Braun, 1996, pp. 23 ff).

To begin implementation of the plan the ICPR deployed more than 150 experts to develop ways to reach clean-up and other goals on schedule. Representatives of each country and the European Commission reaffirmed the need to strengthen regulations on accidents, to list the sites which represented a particular danger to the Rhine, among other specific actions, and have taken or are planning numerous legal, administrative, technical and organizational steps related to safety.

Flooding, Winter of 1994-1995

In February 1995, a combination of unusual natural phenomena compounded by the extensive alterations made over decades to the course of the river caused flooding which resulted in the deaths of at least four people and the temporary relocation of 250,000 Dutch residents. Large areas of cities on the Rhine were submerged (Simonian, 1995). Blame fell primarily on a century of river development which had ignored management precautions.

Beginning in the 1950s the construction of ten hydro-electric plants along the upper Rhine on the Franco-German border essentially created another river, which ran parallel to the old Rhine or used existing branches and travelled at twice the speed of the former river. It was this phenomenon which caused a steady increase in flooding, culminating in the disasters of February 1995. Areas that served the function of absorbing overflows have been cut off the river, replaced by concrete embankments. Remnants of dried marshlands were used first as extra farm land and later for residential and industrial development. Ninety per cent of the river's wetlands have been reclaimed. The need for the construction of additional hydro-electric power plants to supply riparian development further altered the natural course of the river. New branches were deepened and straightened to facilitate shipping so that barges could by-pass the new water project construction through a network of locks and side canals (Simonian, 1995). The result has been to shorten the river by an estimated 100 kms. or more.

A number of proposals have been made to remedy the situation. A natural management method is suggested by George Rast, World Wildlife Fund, Rastatt, Germany. A model flood plain forest can be seen at Rastatt. Flooding is now controlled by engineering technology but Rast hopes that the river will be allowed to take over again soon. Technical solutions include the diversion of the flood waters by special operation of the hydropower plants, construction and operation of retention weirs, and construction and operation of polders. The latter steps were begun in the early 1980s but the project is incomplete.

4.3. Evaluating Rhine River Basin Management

Agreement on a common forum to solve problems and share data

- A common forum for discussion of practical ways to improve the management of the Rhine Basin was in place in the 1950s.
- The impact of extensive development within the basin and the steadily increasing levels of municipal and industrial pollution in the middle and upper reaches of the river have been discussed but issues have been ignored until crises developed.

Binding convention for all countries

- The basic agreement to protect the environment of the basin resulted in the formation of the ICPR, an organization of five basin states and the EC, which has administrative functions. In September of 1996 the six countries agreed to regulate the collection and elimination of waste generated by boats in the river, the result of an initiative by the Central Commission for the Navigation of the Rhine.
- Considerable discretion remains in the hands of the separate states as they proceed from agreement to agreement in response to a number of critical issues. Recent disasters have prompted a closer look at objectives and the need to exert more control over pollution and flooding. Fred Pearce, writing in New Science points out that despite problems "the Rhine is the first major European river where restoration has gone beyond pollution control to a planned ecological revival. The lessons learnt are certain to be adopted elsewhere on the continent, from the Clyde to the Danube (Pearce, 1993a)."

Sharing data:

- Sharing of much data is on a voluntary basis. Increasing pressure by environmentalists has resulted in some public sharing of data within and among countries. Exchange of data for emergency warning purposes has improved.

Basinwide plans:

- Extensive development of the basin for navigation, hydroelectric power, and flood protection has not been unilaterally or bilaterally planned. The action programme sets limited basin-wide protection goals.

Both disasters have made each country, including Switzerland which had not fully participated in the ICPR, more aware than previously of the need to have basinwide environmental impact assessments for all projects. The EC participation in the ICPR is a positive factor in this regard.

PART TWO: The Danube Basin, Pre-Transition

CHAPTER 5

The Danube before Communism

5.1. Basin Profile

In the Danube basin, as in international river basins elsewhere, conflicts may arise from its geography, competition for water use, power balances shifts within and among countries, and historic ethnic or cultural antagonisms. Caught in an interplay of competing empires from its earliest recorded history, the region has endured repeated stress over competition for natural resources related to the river and its tributaries. Power balances in the basin have shifted periodically over the centuries with each significant change leaving an imprint on the population groups which emerged as nation states in the 19th century. Economic and political development was controlled for centuries by investors from outside the basin. Historic and regional rivalries took hold and remain a threat to peace in the region and in the continent. Opportunities for countries to support peaceful cooperation for basin resource management were minimal throughout the pre-transition period.

5.1.1 GEOGRAPHY

Traditionally, pre-1989 commentary on the Danube River introduced it as the second longest river of Europe after the Volga, rising in the Black Forest mountains of west Germany and flowing for approximately 1,770 mis. (2,850 kms.) to its mouth in the Black Sea. The river basin was said to include eight countries (Germany, Austria, Czechoslovakia, Yugoslavia, Hungary, Romania, Bulgaria, USSR) and might have added the four more which shared a very small part of its large catchment area: Switzerland, Italy, Poland and Albania. These four countries occupy less than two percent of the total Danube drainage area and are not included in the Danube Program, Convention or this book. By 1993, the eight had become 13 as Czechoslovakia divided into the Czech Republic and Slovakia, the former Yugoslavian Federation countries of Slovenia, Croatia, and Bosnia declared their independence, in company with the former USSR republics of Moldova and Ukraine.

The Danube is called by that name at the junction of two smaller rivers, the Breg and the Brigach, in the quaint German town of Donauschingen in Baden-Württemberg. Its upper reach extends northeastward through mountains and forests, making its way through an impressive valley between Ingolstadt and Regensburg. It is between these two cities, at Kelheim, that the Main-Danube Canal joins the river, thus connecting the North and Black Seas through the Main, a Rhine tributary. The canal began operation in September 1992. At Regensburg the river turns south and flows through a fertile and flat section of Bavaria, crossing into Austria at the city of Passau. Here it is joined by a major tributary, the Inn. To improve navigation and provide energy, a number of dams and dikes have been built along this reach of the river. The upper Danube, some 600 miles long, has a rapid current of from three to five miles an hour.

The middle course of the Danube extends from what is called the Hungarian Gates near Bratislava to the Iron Gate gorge in the southern Romanian Carpathian Mountains. This stretch of the river is wide, occasionally measuring a mile from bank to bank and somewhat less swift than its upper reach although the rate of flow varies. There are major tributaries entering the river in this stretch, the Drava, the Tisza, and the Sava which have substantial impacts on the river's regime. The Sava joins the Danube at Belgrade, upstream of the Iron Gate reservoir. The average runoff increases substantially from just north of Budapest to the landmark Iron Gate dam project which is jointly operated by Yugoslavia (Serbia) and Romania (National Geographic, 1986, p. 574). One of the largest of the Danube projects, it includes the Djerdap High Dam and the Iron Gate power station. This project includes locks which make navigation through the gorge possible.

From the Iron Gate, the third section of the river forms part of the boundary between Romania and Bulgaria, crossing an extensive and wide plain. Tributaries here include the Olt, the Siret and the Prut. The Danube then heads to the north until it reaches the Romanian city of Galati where it makes a sharp turn eastward. The delta begins at Tulcea, some fifty miles from the sea. Navigation to the Black Sea is through the Sulina channel which was completed in 1980. The delta, rich in bird and other species, has received special attention from international environmental organizations; a substantial part of it has been declared a biosphere preserve. It has, nevertheless, sustained severe losses in resources from pollution and other problems

5.1.2. MISMANAGEMENT PROBLEMS

The Danube Basin shares basic mismanagement problems with other international river basins. Aridity and pollution in its lower reaches have caused losses of large farming areas and reduced the population of threatened communities. Most of its countries must look beyond their borders for a substantial percentage of their water resources. The economic and political infrastructures of the former Communist countries remain under such severe pressure that it is difficult for them to put a high priority on transboundary resource management. The potential for the use of military power to settle historic resource conflicts, vividly apparent in Bosnia, is present throughout the region. Firmly supported basinwide instruments for resource management might well have prevented or modified the development of the Slovak Gabčikovo water project--the large and controversial hydro-electric dam constructed downstream of Bratislava--and other less publicized development problems in basin countries.

5.2. Post-World War I to World War II

The prizes for domination of the Danube by vying power blocs were protection of trade routes from the east to the interior and from the north and the west to warm water ports. "In modern times," the historian Davies summarizes, "[the Danube] catchment area supplied the territorial base for the great multinational empire of the Habsburgs, and the scene for the principal confrontation in Europe between Christianity and Islam (Davies, 1996, p. 63)."

Prior to 1914 the Danube Basin was dominated by Austria-Hungary. Four other countries in the basin, Bosnia, Serbia, Bulgaria and Romania combined did not equal it in size. The following summaries describe the Danube countries as they emerged after the Great War and provide a general idea of their political and economic development during the traumatic period of economic depression and growing fascism. The period between World Wars I and II was marked by political instability in the Balkans and in all of the Danube states.

5.2.1. DECLINE OF EMPIRES

The demise of four empires, the Russian, German, Turkish, and Austro-Hungarian, left Eastern Europe temporarily without external oppressors. The small states that emerged from peace treaties signed in 1919-1920 diverged in political and economic approaches to their problems. Fledgling democracies lacked support that might have come from the west if the World War I victors had themselves been stronger. As it was, the power vacuum within Germany was soon filled with a brutal dictatorship which sought allies among the conservative anti-semites of the fragmented Danube countries, especially Austria, Hungary, Romania, and Bulgaria.

The initial post-World War I hardships, unemployment, inflation, a devastating influenza epidemic, and severe political unrest, were followed by the depression of the 1930s. Ripe for the courtship of Nazi Germany, most of the Danube countries fell victim again to supporting the losing side in a World War.

5.2.2. KEY TO THE FUTURE

The economic and political turmoil of this period in eastern Europe is key to an understanding of its communist future and its response to the opportunities of the post-Communist transition period. The goals of developing power blocs in west and east in the early 1920s were essentially what they had been in the past. Germany sought lebensraum and where else to find it but through the access provided by a weakened Austria, and the new Hungary, and Czechoslovakia? The post-revolution Union of

Soviet Socialist Republics shared the warm water goals of its forebears and cultivated the national aspirations of Slavic peoples. The plentiful resources of the Danube Basin: agricultural, mineral, and to some extent industrial, augmented by a convenient transportation route and opportunities for energy development were an easy target when the countries which owned them were in economic and political chaos.

5.2.3. PROBLEMS BY COUNTRY

The peace treaties that followed World War I determined new boundaries for most of the Danube countries. Austria, stripped of the extensive territory it had shared as part of the Austro-Hungary Empire, became a small country of six million. Its population suffered acutely as the result of post-war problems which included starvation as well as inflation and unemployment. Political unrest led to riots in the 1920s and the establishment of a dictatorship. Its chancellor, Engelbert Dolfuss, attempted a balance between the political left and right and was assassinated by Nazis in 1934. Four years later, in March of 1938, Austria was occupied by German troops.

Czechoslovakia

Czechoslovakia was founded in 1918 after the breakup of the Austro-Hungarian Empire. The Treaty of Saint-Germain recognized the new republic of Czechoslovakia which had been proclaimed by its leaders the year following the joining of the Czech and Slovak portions of the former Austro-Hungarian Empire. The political and economic future of the new country seemed assured, including as it did the major portion of the industries of the former Empire. Its leaders immediately sought to make life better for peasant farmers by instituting a land distribution program. They coped with the depression of the 1930s, preserving a democratic system with the help of an enlightened citizenry (Halasz, 1959, p. 211-2). Its leaders, Presidents Tomas Masaryk (1918-1935) and Eduard Beneš (1935-1938), tried to find friendship and protection in western Europe, particularly in France, and formed a Little Entente with Yugoslavian and Romanian leaders.

The centralized government faced internal ethnic problems from the start. There were significant economic and cultural differences between the Czechs and Slovaks. A large German minority joined in agitation against the central government. From the early 1930s the German leader Adolph Hitler backed its request for annexation with Germany. The Munich pact signed by Germany, Great Britain, France and Italy in September 1938 provided for the cession of the Bohemian border lands (Sudetenland) and other parts of Czechoslovakia to Hungary and Poland. The new state was renamed Czecho-Slovakia in November 1938 and divided into three autonomous units, Bohemia and Morava; Slovakia, and Ruthenia (in western Ukraine). Benes resigned the presidency and shortly thereafter Hitler forced his successor to make Bohemia and Morava a German protectorate and to give Ruthenia to Hungary. Slovakia was, at least in name, independent.

After the outbreak of war Benes established a provisional government in London. Refugees from the dismembered country fought with Allied forces throughout the war.

In April of 1944 American forces entered from the west and Soviet forces, accompanied by a Czech military force and Benes and many of the leaders who had served the provisional government, entered from the east.

Hungary
The Treaty of Trianon (1920) stripped Hungary of more than a third of its territory and reduced it to a size close to its present boundaries. Political chaos followed. An initial government set up in March of 1919 under Michael Karolyi proclaimed the country a republic but collapsed in a few months as the communist and socialist parties under the leadership of Bela Kun seized power. The so-called "red terror," which included an invasion of Slovakia, was followed by a "white" terror when Romanian forces intervened and defeated Kun's forces. French troops, under the leadership of Admiral Miklos Horthy de Nagubanya, participated in the overthrow of Kun. A provisional government proclaimed Hungary a monarchy. No king ever served; Admiral Horthy was designated regent and signed the Treaty of Trianon. He remained in power until near the end of World War II.

Horthy pursued domestic policies which encouraged anti-semitism and protected feudal landowners. Hungary had lost almost all the land occupied by non-Magyars and valuable natural resources. As its efforts to regain this land were frustrated by the Little Entente leaders and France, it became allied with fascist Italy and Nazi Germany.

As a reward, Hungary regained territories from Czechoslovakia, Yugoslavia, and Romania, holding them from 1938 to 1944. The Russians invaded in 1944 and, after taking eastern Hungary, surrounded Budapest in December 1944. Subsequent fighting between German-Hungarian defenders and Soviet forces destroyed large parts of Budapest.

Horthy had first sought a separate armistice with the Russians in October 1944 and then resigned to be replaced by an ultra-right group called the Arrow Cross. The temporary Hungarian government signed an armistice with the Allies in January 1945 which provided for the return of territory it had gained under the Nazis to Yugoslavia, Romania, and Czechoslovakia as well as the payment of substantial reparations to the Soviet Union..

Yugoslavia
Created as a new state in 1918, Yugoslavia (initially called the Kingdom of the Serbs, Croats and Slovenes) included the independent countries of Serbia and Montenegro and the former Austro-Hungarian areas of Croatia, Slovenia, and Bosnia and Herzegovina. In the face of moves by the neighboring countries of Bulgaria and Hungary to regain territory lost under the World War I peace treaties, Yugoslavia entered into alliances with the Little Entente and France. Internal problems continued. Serbian and Croatian leaders set up a separate parliament in 1928 in Zagreb, and King Alexander proclaimed a dictatorship and changed the name of the Kingdom to Yugoslavia. Alexander was assassinated in 1934 in Marseilles by a Ustashan, a member of a Croat fascist organization. Prince Paul, the heir apparent, gradually allied with the Axis powers

In 1939 the Belgrade government gave autonomy to Croatia and in 1941 Prince Paul made a nonaggression pact with Hitler which did not require Yugoslavia to enter the war. The response by Serbian leaders was to seize power, exile Prince Paul, who had been acting as regent, and install King Peter. In April 1941 Hitler bombed Belgrade and German, Italian, Hungarian, and Bulgarian troops invaded the country. Less than two weeks later the Yugoslav army surrendered.

Hitler and Mussolini dismembered Yugoslavia, giving the greater part of Slovenia to Austria and leaving the coastal lands and Montenegro to Mussolini. Bulgaria occupied Serbian Macedonia, Albania annexed the Kosovo region in south Serbia and Hungary took over the Bachka and Barana; Croatia became independent, enlarged with the annexation of Bosnia-Herzegovina with its Muslim and Serbian population while Serbia itself remained independent. Fearful atrocities were committed by the Axis occupation and their allies.

The government in exile established by King Peter remained in London but many guerrillas continued to fight the invaders. Partisans under Tito (Josef Broz) were organized as an army in Bosnia and by late 1944 had expelled the Germans from Yugoslavia. After national elections in November 1945 Tito was proclaimed the head of a new republic, a federation of six "equal" units: Serbia, Croatia, Slovenia, Macedonia, Montenegro and Bosnia-Herzegovina.

Romania

The monarchy established in Romania in 1881, after the country had achieved full independence from the Ottoman Empire, did not end until 1947 when the Communists came to power. Monarchical longevity belied the chaos which the country endured during this period. The Romanians, who had entered the war on the Allied side in 1916, ultimately shared in the territorial prizes bestowed by the Treaties of Saint-Germain and Trianon. Romania's area doubled to 296,400 sq. kms. (114,000 sq. mis.) and included the regained territories of Transylvania and Bessarabia. Reforms were instituted after the war, in particular, the break-up of the large estates and the distribution of land to the peasants.

Domestic problems were acute, however, as large Maygar and other minority enclaves fought Romanian rule. Democratization of the political system existed in form only as electoral laws were constantly violated and repression of civil liberties rampant. Economically, Romania remained undeveloped. In 1930 the previously abdicated King Carol returned and established a court dictatorship which was universally unpopular. The opposition was split among left and right groups, with the latter eventually gaining the upper hand. It was allied with extreme right-wing, anti-semitic organizations which were linked with a terrorist organization called the Iron Guard. A relatively liberal prime minister, Nicolas Titulescu, was forced to resign in 1936. King Carol II proclaimed the rule of a single political party, the "Front of National Renascence" in 1938. In 1940 Romania became a neutral Axis ally. Fascists took over the government in 1941 and Romania joined the Germans in the attack on Russia.

Romanian forces in Stalingrad suffered severe defeats. Generally on the defensive in the period of Red Army resurgence, Romania was occupied by Soviet troops in August 1944. A coup d'etat by King Michael took over from the fascist regime and Romania entered the war on the Allied side. A peace treaty between the Allies and Romania signed in Paris in 1947 confirmed the armistice agreed to in 1944. Romania recovered all of its territories except Bessarabia, Northern Bukovina and Southern Dobruja.

Bulgaria
Bulgaria, on the losing side in World War I, was ruled by King Boris III until his death in 1943. The treaty of Neuilly cost the country its outlet to the Aegean Sea which was given to Greece; it also lost territory to Yugoslavia. Reparations which had to be paid to the victors, the influx of many refugees from the lost territories, and severe economic problems kept Bulgaria in turmoil after the war. Political democratization had about the same level of success as it did in Romania. The agrarian party in power, under the leadership of Stamboliski, was overthrown by a military coup in 1923. Thereafter, the political arena was dominated by Macedonian terrorists until 1934 when the army installed a new prime minister. Shortly thereafter Boris III established his own dictatorship. Bulgaria joined the axis in 1941, declaring war on the US and England but not the Soviet Union.

Boris's death in 1943 left the country without a strong leader, and in September 1944 the Soviet Union declared war on Bulgaria and quickly invaded the country. Opposition forces, the Communist and Agrarian parties as well as Army officers who sided with the Soviet Union, seized power. Bulgaria belatedly declared war on Germany and an armistice followed.

5.3. The Aftermath of World War I

Power blocs which had played see-saw matches over east European countries for decades were vastly weakened by World War I. Their return to full strength was expected to be a slow process. The reconstituted countries lining the banks of the river and subsisting on its resources wrestled with problems engendered by ethnic rivalries which had been worsened by war, their own unresponsive governments, and the severe economic problems which prevailed throughout Europe. While some countries enjoyed a higher level of economic development and democracy than others, overall the seeds of fascism landed in friendly soil.

The economic and political status of each east European country in the basin from the new states of Czechoslovakia and Yugoslavia to the reconstituted countries of Hungary, Bulgaria, and Romania varied in the period before World War II. Czechoslovakia had a developed industrial economy and democratic institutions. One or two of the Yugoslavian republics had become partially industrialized. Hungary had grown commercially and intellectually in the second half of the 19th century. Increasing farm productivity and industrialization depended on the ability to use foreign capital productively. In a manner common to developing countries everywhere, foreign investment for the most part, in the view of one Czech journalist, "went into colonial exploitation of raw materials not integrated into the framework of the domestic economy, or else into armaments. Foreign aid in capital and 'know how' to turn potential into actual wealth was outside the scope of the interwar era (Halasz, Nicolas, p. 76)."

World War II brought changes to the governments and economies of the Danube east European countries. The industries of Czechoslovakia and Hungary geared up to produce weaponry for the Wehrmacht. Romania provided agricultural produce and petroleum. Except for Yugoslavia governments were lead by Nazi puppets and vestiges of democratic institutions all but stamped out. War damage to municipal and industrial infrastructures was extensive. In any case, significant decisions about the future of the Danube basin countries continued to be made for them by policy makers outside the basin.

CHAPTER 6

Political and Economic Issues under Communism

6.1. Impact on Environmental Policies

The Soviet presence in the Danube east European countries from 1945 to 1989 significantly affected the ability and willingness of basin countries to agree formally or informally on a basin-wide approach to economic development and natural resource protection. The Cold War prevented the adoption of such necessary steps in basin management as the free sharing of economic and environmental data or the enforcement of common water quality standards.

While EC council directives set the pace for pollution control for western countries and strong environmental movements generated support for it in the parliaments of several countries, eastern Europe adopted Soviet style legislation: a plethora of standards for a very large number of parameters with little enforcement and very low investments in waste treatment facilities. Trade and other forms of cooperation were encouraged by the Council for Mutual Economic Assistance (CMEA), but divisive Soviet policies and deteriorating economies made a comprehensive pact on Danube basin management impossible.

Concern that the air and water western Europe shared with the east might lead to serious health crises grew in the 1970s and 1980s. International commitments to wetlands protection, reduction of harmful air and water emissions, and the like, were made by eastern countries but carried little weight. Treaties with a broad focus lacked the enforcement potential of a binding agreement among all the Danube countries to adopt compatible environmental legislation and enforcement procedures.

6.2. East and West Divide

Shortly after the unconditional surrender of Germany in May 1945 it was clear that serious divisions about the future of Central and Eastern Europe prevailed among the wartime allies. Soviet leaders were determined to exert a tight control over their neighbors. The failure to resolve the key question of reuniting the four zones of occupation (United States, British, French, and Soviet) in Germany and the general vagueness of the agreements reached at the Postdate Conference held in July 1945 contributed to the final break-up of the World War II "Grand Alliance." The USSR began unopposed military and political strategic moves which ended in its domination of all the east European Danube countries. Austria and Germany eventually adapted to the political and economic ways of the west, received substantial financial help, and prospered.

6.2.1. GERMANY AND AUSTRIA

In 1949, shortly after the western powers permitted their separate zones in Germany to unite and restore parliamentary democracy in a new Federal Republic of Germany (FRG), the Soviets installed a communist regime in the eastern section, calling it the German Democratic Republic (GDR). Industrialization in the FRG proceeded at a brisk pace. The corporate infrastructure of the past was put to use, including the giant cartels which were supposed to be banned but were not. They, and the astute and firm leadership of Konrad Adenauer, launched a remarkable economic recovery.

In 1945, Austria was reestablished within its prewar boundaries and divided into four administrative zones for occupation by U.S., Soviet, British, and French forces. The four powers did not agree to a treaty until 1955. A series of socialist governments helped to bring prosperity to Austria which profited much from western help. The country's basic industries were nationalized, and the Austrian economy became increasingly industrialized.

Despite the small share of land which is arable (20% of the total) and a decrease in small-scale farming, scientific methods and consolidation of agricultural resources have made Austria almost self-sufficient in food.

6.2.2. EAST EUROPEAN COUNTRIES, CONSTITUTIONS AND GOVERNMENTS

In the east European countries, meanwhile, the citizens of the Soviet-style governments were having somewhat different experiences. Domination by the Soviet Union choked off the few commitments to democracy which had been nurtured in Czechoslovakia and in parts of Yugoslavia. Agriculture waned and countries that had been self-sufficient in food or even exported produce could no longer feed their own populations. The vaunted goals of industrialization were achieved piecemeal, with "specialization" enforced so that no one country could make unilateral economic gains. While the relatively new issue of environmental protection of natural resources, including those in the Danube River Basin, was on the communist agenda, standards were not enforced and few investments committed to waste treatment management. Each country's experience differed to a degree. Such differences inevitably affected the political, economic, and environmental policies in the post-communist period.

During the period of Soviet domination, east European countries lost democratic institutions or replaced one form of autocracy with another. Conflicts developed not between countries but within them as discontent erupted during the 1950s in Poland, East Germany, Czechoslovakia, and Hungary.

Soviet leaders, considering themselves justified for a number of reasons, insisted on friendly governments in eastern Europe in the aftermath of World War II. In practice, this meant the installation of Soviet-style regimes in all the east European countries under Red Army control. By 1949 all of the east European Danube countries had some variation of the USSR constitution of 1936 in place.

Soviet Model

The Soviet model rested politically on the "dictatorship of the proletariat" and the public ownership of land and means of production. Legislative power was in the hands of a supreme soviet or assembly. Constitutionally, executive power was vested in the council of ministers or commissars appointed by the supreme soviet; its chairman was considered the premier. The central committee of the communist party made policy decisions and the first secretary of the presidium of the central committee was ipso facto the actual head of the state. The judiciary served the commands of the party and protected the socialist state by educating citizens in such socialist principles as the maintenance of labor discipline, achievement of production quotas, informing authorities about hostile acts of individuals, and the like.

Variations from the Model

The variations in political practices in the five Danube east European countries under communism are probably more interesting and enlightening than their similarities. Soviet leaders regarded each country differently, for historic, economic, and political reasons. The new Communist leaders within each country manipulated the Soviet government format to suit their tastes, most often with more regard for their own power positions than the needs of the general public. These variations are summarized below, following the position of the countries from the upstream country of Czechoslovakia to the downstream country of Bulgaria.

Czechoslovakia

In Czechoslovakia before the second world war Tomas Masaryk, president of the new republic from 1918 to 1935, and Edward Benes who succeeded him as president, had contributed to the development of democratic institutions. Benes returned from exile after the war and joined a coalition government elected in 1946 with the Communist Party as the leading party. In February 1948 the Communists took complete control of the government and the first Soviet type constitution was adopted. Benes resigned from the Cabinet. The unicameral legislature passed a number of measures to meet constitution goals, including rapid industrialization, nationalization of business and industrial organizations and collectivization of agriculture. Early in this period Czechoslovakia became a member of the UN as well as the Soviet Union's trade organization, the Council for Mutual Economic Assistance (CMEA), and the Warsaw Pact.

In 1968, during a very brief time called the "Prague Spring," Alexander Dubcek became the Communist Party first secretary and attempted to introduce democratic reforms. The move to formulate democratic institutions was stamped out when the country was invaded by Warsaw Pact forces and Dubcek deposed. Subsequently a constitutional law in 1969 federalized the country, creating, in effect, two separate socialist republics, each with a number of delegated powers. The federal government kept authority over foreign policy, defense, and natural resources. There was joint control over planning, currency, prices, and domestic industrial, agriculture and other policies including the police. In January 1977 over 300 individual professionals signed "Charter 77," a petition to government leaders to grant constitutional rights, support various UN covenants, and in other ways ameliorate the oppressive communist rule. Their activity was tolerated but had little effect. The control which leader Gustav Husak had reaffirmed and tightened in May 1975 remained in effect.

The level of economic development in Czechoslovakia after World War I had been the highest in central and eastern Europe, and had matched that of some countries of western Europe. With the advent of Communism and the imposition of Soviet economic practices, a new era began. Farms were collectivized, central economic planning established and heavy industry, mining, and manufacturing emphasized. Eventually, in the late sixties, with a relaxation of the more stringent policies of the early period, successes in economic output tended to balance weaknesses in other areas, levels of labor productivity, energy output and agricultural production. Over all, the Czech contribution to the entire eastern bloc and to the Soviet Union was substantial. It supplied them with complete manufacturing plants and, of special value to the Russians, armaments.

Agriculture, although almost entirely collectivized, performed fairly well in the last years of Communist rule, in part because it had begun to receive a larger share of all new investment. Aging equipment, outmoded industrial plants, and general shortages in raw materials and other resources hampered the development of manufacturing and other branches of industry. Transportation and communications, although advanced in comparison with the rest of eastern Europe, were inadequate. All of these factors, combined with the negative economic trends of the 1970s, led to a decline in economic performance, and the standard of living stagnated during the late 1970s and 1980s.

While in many ways it fit the Soviet economic model, Czechoslovakia had a per capital GDP in 1989 of $8,580 which put it at the same level as Spain or Ireland (Kaiser, 1994, p.507). The need for reform to improve environmental protection was recognized but little was done to resolve persistent problems. In particular, the health dangers of using soft coal for energy were recognized but nothing done to curtail their impact.

It depended on the Soviet Union for petroleum and raw materials, especially iron ore. Forty per cent of its trade was with the Soviet Union and other EE countries. It also joined the Soviet Union and other socialist countries in providing economic and technical assistance to third world countries. Its gross debt to the west grew but was not as high as other bloc countries.

Hungary

Hungary, which had been occupied by Soviet forces at the end of the war, agreed to a treaty in Paris in 1947 which reduced it to the boundaries set after World War I and required the payment of $300 million in reparations to the USSR, Czechoslovakia, and Yugoslavia. An elected coalition government began land and other reforms but was taken over by a Hungarian communist party coup early in 1948. The country became a "people's democracy" under a Soviet style constitution in 1949. Considerable discontent during the next few years led to a popular uprising in 1956 which was overcome with the use of Soviet troops and tanks. Reforms instituted by Imre Nagy, premier from 1953 to 1955, deemphasizing intensive development of heavy industry and the collectivization of agriculture, and relaxing police control, came to an end. An estimated 100,000 supporters of these reforms left the country after the defeat of the uprising. Nagy himself was executed by order of a Hungarian court in 1958.

A few years of repression followed but then Janos Kadar, who was sympathetic to and favored by the Russians, took over as head of the government and began a series of economic reforms which stabilized the country politically and economically. Hungary also became a member of the UN and the Warsaw Pact. Kadar remained as head of the government until 1988 when he was replaced as general secretary of the party by Karoly Grosz.

The Hungarian communist economic mode was perhaps the most pragmatic of all the eastern bloc countries, particularly after the 1956 rebellion and the development of "goulash Communism" under Kadar. The country received help from an alarmed Soviet leadership and from other eastern European states in significant amounts for recovery from damage to physical assets, loss of skilled technicians, and other legacies of the armed intervention. When it was evident in the late 1950s and early 1960s that emphasis on heavy industry had stifled production of consumer goods and inflicted much ecological damage, a new program called the "New Economic Mechanism," adopted in 1968, resulted in a partial market-oriented restructuring of the economy. The most important policy was to allow "greater scope for private initiatives, first in agriculture and then in the other sectors (Marer, 1994, p. 483)."

Initially successful after an infusion of billions of dollars in foreign loans, the economy nevertheless faltered in the 1980s. It has been estimated that reforms had moved the country about one-third of the way toward a market economy. In 1988 the failure to reform the industrial sector development so that resources would be protected brought about a major economic crisis, which helped to overthrow the socialist system.

Yugoslavia

Josep Broz, who had become famous during World War II as Tito, the leader whose partisan guerrillas had defeated German troops and forced their withdrawal from the country, established a de facto government in Yugoslavia by 1943. The rival anti-Nazi organization, the Chetniks, were denied participation in the government and the king-in-exile not permitted to return. A Soviet style constitution was adopted in 1946. Tito almost immediately, however, displayed disloyalty to and questioned Soviet policies. In retaliation the "Cominform" bureaus established in each satellite country by Stalin in 1947 adopted a resolution expelling the Yugoslav Communist Party the following year (Staar, 1988, p. 224). "National Communism" was adopted by Tito while he maintained independent foreign and economic policies.

The six federal republics and two autonomous provinces, Kosovo and Vojvodina, were under the control of the central government with respect to most important matters. Constitutional changes were made in January 1953, in 1963 and in 1974.

Basic law, Yugoslavian style, was the "longest and most detailed of all bloc constitutions in its definition of social system and government (Staar, 1988, p. 234)." It also transferred some power to the republics." In reality, though, the country's political life belonged to Tito and the Yugoslav National Army (Pesic, 1996, p. 13)."Tito served as party leader and the president of the federation until his death in 1980. With effective domestic policies and strong emphasis on continued military strength, Tito did not need to depend on Soviet military and economic support for his leadership. He adopted a policy of nonalignment with either east or west although he attempted to keep a foot in both camps.

Yugoslavia's economic development initially followed eastern bloc principles but rigid planning in agriculture and industry soon gave way to a combination of more flexible central planning and private initiative. Farm collectivization was abandoned in 1951 and most agriculture in Yugoslavia was thereafter performed by peasant land holders. Intensive efforts to industrialize changed an agrarian economy into a mixture so that the largest share of the work force was employed in the industrial and commercial sector. The break with Stalin prompted the west to provide increased financial assistance. Attempts to standardize the economic level of all the republics, however, aroused antagonism between the "haves" (Slovenia and Croatia) and the remaining "have-nots." "Between 20 and 30% of the Slovene gross national product [in 1988 was] diverted to southern republics, which [caused] resentment (Staar, 1988, p. 230)." Resulting tensions in Croatia and elsewhere were a taste of things to come in the early 1990s.

Yugoslavia was hit hard by inflation and a lowered standard of living during the 1970s and 1980s. Several economic indicators fell sharply in the 1980s, as regional disparities in economic development and structured imbalances induced by socialist institutions contributed to a general economic breakdown. After Tito's death the central government weakened, and separatist movements developed in the federation. In addition to the resentment shared by the more economically advanced countries, long-standing ethnic rivalries revived.

Romania

In Romania Soviet occupation force leaders toward the end of the war supported the installation of a modified communist regime. By December 1947 King Michael had been deposed and Romania proclaimed a people's republic, with Gheorghe Gheorghiu-Dej as its new leader. A succession of constitutions in 1948, 1952, and 1965 confirmed the Soviet style of rule, at least on paper. In 1965, a revised constitution put Romania on track with the other east European countries and it became known as the Socialist Republic of Romania.

Nicolae Ceausescu, who initiated a takeover of the Communist Party and the government when Gheorghiu-Dej died in 1965, emphasized independence in political and economic policies. Romania sought western contacts while remaining a member of CMEA and the Warsaw Pact. Beginning in the mid-1970s Ceausescu became obsessed with identifying all achievements as the product of his own leadership capabilities, ignoring criticism and rejecting accurate appraisals of very low levels of agricultural and industrial production. His dictatorship lasted for over twenty years, becoming increasingly arbitrary and oppressive, until his death by assassination in 1989 following a popular uprising.

The occupation of Romania by Soviet forces during the latter part of World War II had an unfavorable impact on the Romanian economy. The country paid reparations of $300 million in goods (at 1938 prices) and also provided petroleum products for much lower prices than would have been received on the world market (Staar, 1988, p. 200). During the communist era industry was nationalized, and heavy industry, particularly machine-building, received the major share of state investments. During the 1970s and 1980s investments yielded relatively low increases in productivity, and aging fixed industrial assets became increasingly energy inefficient.

During the communist era more than 90% of Romania's agriculturally productive land was collectivized into state and cooperative farms. Agriculture employed more than a quarter of the work force, but farm productivity was low due to lack of incentives, ill-advised investment priorities, bad management, and under-mechanization. Although domestic food shortages characterized the last years of the communist regime, Romania remained a net food exporter.

Bulgaria

Defeat by the Germans and Bulgaria's subsequent occupation by Soviet forces in 1944 paved the way for local communist supporters to take over the Bulgarian government. Stalinization took hold in Bulgaria perhaps more aggressively than in any other East European country. Soon, large numbers of persons who were considered threats to the regime were being tried as fascists or traitors and summarily executed (Staar, 1988, p. 37). In 1945 the Bulgarian communist Georgi Dimitrov, who had lived in the Soviet Union for the duration of the war, returned to Sofia and briefly served as the head of the newly installed communist government. He died in 1949. Todor Zhivkov was elected party leader in 1954 and was subsequently re-elected seven times. He was prime minister from 1962 to 1971 and in that year also became head of state. By 1971 Bulgaria's leaders determined that the new stage of communism, the "socialist state" phase, had been reached and a new Soviet-style constitution was adopted.

The Bulgarian communist leaders were perhaps the most devoted to the dogma and practice of the Soviet Union. Ironically, the "orthodox Communists were so anxious to imitate the USSR that they found themselves making the same mistakes long after the Soviets had taken a new approach and attempted to rectify their errors (Staar, 1988, p. 38)." Bulgaria received substantial economic aid from the Soviet Union in return for continued loyalty, a steady relationship which was appreciated during the defections of Yugoslavia and other bloc countries. While estimates vary, Bulgaria received "a disproportionately large share of Soviet financial support (Brown, 1988, p. 320)." Energy supplies, especially oil, was provided at soft currency rates and Bulgaria permitted to refine it and sell it on world markets.

Bulgaria had little modern industry before World War II. The Soviet Union supported a modest development of heavy industry and the manufacture of electronic equipment including computers and electrical products such as fork-lift trucks (Brown,

1988, p. 324). An early member of CMEA, Bulgaria traded primarily within the eastern bloc. While official government statistics recorded high industrial growth rates and consumption standards, downward trends belied the rosy picture. When it became apparent during the 1960s that the early growth strategies were ineffective, a number of reforms were attempted but did not succeed. Generally, the failure to encourage farming in the more pragmatic way of the other eastern bloc countries, plus heavy emphasis on agricultural reinvestment, meant a low standard of living in both urban and rural areas.

6.3. Danube Agreement Status, 1989

It comes as no surprise that a positive, working, agreement to protect Danube ecological resources was not adopted during the Cold War. Many factors inhibited support within and among the countries for such a move. First of all, a forum for discussion was lacking. When east and west government and research specialists met they were usually well chaperoned. The display of support for dialogue over basin environmental quality issues was enough for eastern countries.

When it may be a criminal offense to share environmental data within the country or treason to share it outside the country, one could not expect a general exchange of data which might be helpful to the resolution of international issues. Added to that problem was one created by technological disparities. Some country standards might be close to the west so that justifiable comparisons in trend data could be made. For the most part technology was an estimated 15 years behind that of western Europe, making it impossible to have an effective working agreement on resource protection.

The Gabcikovo Dam project will survive as a monument to the lack of environmental impact assessment for water projects which would affect the Danube basin-wide. The absence of environmental impact assessments (EIA) and the increasing environmental problems of the late 1980s became spurs to adoption of a Danube convention and activities program.

CHAPTER 7

Environmental Policies in the Danube Region, 1947 - 1989

7.1. Need for New Policies

Rivers clogged with sewerage, dying wetlands, lost plant and animal species, and other evidence of the toll the post-war boom had taken of their natural resources, pressured the industrial democracies of Europe to support vigorous environmental enforcement policies in the 1960s and 1970s. The need for transboundary control was inescapable in the air and water corridors of Europe. EU countries adopted stronger environmental legislation in the 1970s, aided by cooperative cross-boundary research which helped to develop monitoring and treatment technology. At the same time CMEA provided a facade of economic and pro-environment cooperation among the East European countries.

All participating socialist countries passed similar legislative controls on air and water pollution and general protective measures for natural resources, but pollution increased as economies deteriorated. Compliance with standards in the two western countries of Austria and Germany was generally high. In the other Danube European countries it was low as a result of inadequate treatment facilities, lack of resources, lack of enforcement and outdated production processes.

7.2. Austria and Germany

7.2.1. ENVIRONMENTAL POLICIES

The two Danube countries of Austria and Germany adopted similar environmental management policies in the decades following World War II. Much of the early post-war industrial boom in western Europe was not immediately accompanied by strict protective measures. The discovery of the impact of pollutants on both air and water in the 1950s and 1960s (including, for example, extensive surveys of the almost moribund state of the Rhine River and other transboundary water resources) generated an environmental movement in western Europe, especially in Germany, the Netherlands and Denmark.

The latter countries led the way in the promulgation and adoption of common environmental standards and their enforcement. Five-year environmental action programs were begun by the EC in 1973 and continued into the 1990s. National clean water and other environmental legislation in each member country was eventually augmented and guided by EC directives. Despite some disagreement on environmental questions, "...consensus had been obtainable hundreds of times across a large number of specific and controversial issue areas (Sbragia, 1996, p. 254) ..."

7.2.2. WATER QUALITY

Water quality legislation enacted in EU countries prescribed standards of potential polluting substances for uses (drinking water, bathing, fish life) in both surface and ground water (for drinking water) and provided for their enforcement. Standards determined classification schemes for uses, specifying levels with the highest rating which could be used for drinking water without treatment or with minimal treatment, and identifying decreasing levels of purity to the lowest, extremely polluted level. Countries were expected to allocate clean water goals to bodies of water and to publicize the degree to which such goals had been reached. Emission limit values were set for the release of pollutants from point or non point sources and discharge permit systems established for their control.

By 1990 the entire population of Germany was served by sewage treatment works, the only western European country other than the UK to have reached that goal at that time. Federal legislation in Germany provided that effluents from industries which contain dangerous substances must be treated before discharge by the "best available technology." Minimum emission levels were established for inclusive categories of industries. Austrian legislation provided for the classification of surface waters by use by 1959. Legislation enacted in 1987 established a fund to subsidize the construction of water supply and waste water treatment projects.

Some discrepancies in enforcement were noted in a 1993 report on pollution in Germany. Discharges into the river Emscher, for example, which includes the waste of a large number of industrial plants in the Ruhr, receives primary treatment only. Such treatment is for biodegradable substances and it is unlikely that all industries discharging to the river meet specific industrial standards. While the rivers discharging to the North and Baltic Seas require the installation of nutrient removal at many sewage treatment works, the Danube has not been so designated by the two responsible states of Bavaria and Baden-Württemberg (WRc, 1993, p. 74).

In Austria very few industrial companies treated effluents before discharge prior to the passage of the Water Act of 1990. There were several rivers which fell below class II, the minimum requirement established set by the new legislation for all rivers. There were also serious reductions of flow caused by the construction of hydropower plant stations upstream of Vienna.

7.3. East European Countries

Intensive industrial development and wasteful farming methods in the socialist countries during the pre-transition period led to increasing degradation of water and other resources. Encouraged through CMEA to follow Soviet leadership in this aspect of of domestic policy, by the end of the 1970s all Danube countries under communist control had enacted environmental legislation which legally established water quality standards, among other environmentally protective goals. Their policies were ineffective for a number of reasons.

7.3.1. WEAK ENVIRONMENTAL POLICIES

Eastern bloc countries, like the Soviet Union, failed in general to allocate resources to waste treatment. Controlled economies lacked incentives to improve manufacturing processes and to reduce waste. While government-funded non-governmental organizations were publicized as contributing to policies of environmental protection, public participation was generally absent until the last years of the communist period. The Soviet style was to identify a very large number of parameters for regulation and allocate levels for water use which were often more strict than those adopted in the west. Standards totalled into the hundreds while monitoring of their level in surface water was frequently haphazard and data reporting, when it occurred, a confidential matter.

Framework legislation customarily preceded specific laws and regulations covering water quality protection and other types of media management. Legislation was passed in Bulgaria in the 1960s which identified air, water, soils, natural vegetation and animals for protection as well as the proposed levels of conservation, rehabilitation, rational use and protection of natural resources. Hungary also passed framework legislation in 1964, which regulated discharges to water ways with a complex set of classifications. Romanian framework legislation was initiated in 1973, with a specific water law regulating use and protection against pollution.

7.3.2. YUGOSLAVIAN POLICIES

In 1985 the Organisation for Economic Co-operation and Development (OECD), at Yugoslavia's request, prepared a study of the country's environmental legislation (OECD, 1985, Chapter 1). The Yugoslavian constitution of 1946 had endorsed preservation of the environment as a universal human right. It also encouraged individual citizens to express themselves about environmental protection through organizations. Cooperative mining and other enterprises were urged to play roles in waste prevention. As an indication of its public concern for environmental protection, by 1985 Yugoslavia had signed over 30 out of 108 international environmental agreements.

The OECD report provides background material on the relevance of economic conditions in Yugoslavia to high pollution levels. A high rate of growth during the 1970s had been followed by a severe recession. Foreign debt was so high that repayment took eight per cent of the national income. A 70% inflation rate was accompanied by a fall in the standard of living of over one-third during the first half of

1980. There were severe problems meeting World Bank standards for environmental protection that were part of loan agreements for industrial and other projects.

The study found that growth had brought with it polluted rivers and an increasing number of waste dumps. There were administrative problems which accompanied the move toward adequate resource protection. Yugoslavia had cross-boundary restrictions, for example, even at the local, commune level which required that local entities sign agreements before water treatment could be effected. No one administrative agency at the federal level had responsibility for water management. Another problem was created by disparities in development among the republics which hindered adequate treatment planning.

Water-management legislation was passed by the federal government in 1974 and 1976. Administrative orders which classified waters by quality and use and specified the maximum permitted concentrations of radionuclides and noxious substances followed. Drinking water standards were established in 1980. As in the other east European Danube countries, the federal legislation lacked effectiveness. Investment programs did not include funding for the construction or renewal of sewerage systems and treatment plants (OECD, 1985, p. 95). In sum, although the environmental legislation of the republics covered from 200 to 275 parameters, actual monitoring of levels and the enforcement of standards was very limited. When the water quality standards of Yugoslavia (the Serb Republic) were entered into the information system INFODANUBE in 1991 standards had been established for 275 parameters (Murphy, 1993).

7.3.3. FAILURES ELSEWHERE

Other east European countries had similar problems with the enforcement of waste control. Examples abound. In the Slovak Republic of Czechoslovakia several hundred exemptions from its Water Act had been approved between 1980 and 1989, primarily for public sewage systems which included those in the city of Bratislava. Economic enterprises did not bother to obtain licenses or attempt to reach required limits.

In Hungary, extensive monitoring showed a sharp deterioration in water quality over a thirty-year period (late 1950s to 1985). The number of streams registering poor water quality and continued deterioration had increased significantly, particularly in the Tisza River, a major Danube tributary (Hock, et al, 1990, p. 76). Monitoring procedures were inadequate and exceptions constantly made for less stringent special limit values.

Romania and Bulgaria also experienced high levels of pollution. Only a small fraction of water discharged to receiving water from industrial plants in urban areas were treated adequately. Non-compliance reflects severe shortages of treatment facilities as well as inadequate production and inefficient operating practices.

7.4. Danube Basin Cooperation

Soviet domination kept potential political, economic, and ethnic conflicts among the east European Danube countries under control. Allegiance to Moscow by most, if not all, of the country leaders, to the Warsaw Treaty Organization, and to economic ties under CMEA contributed to regional cooperation. Agreement on the shared use of the Danube River Basin was limited to navigation regulations despite suggestions that the mandate of the Danube Commission be broadened to include pollution control.

Growing recognition of the danger of transboundary pollution led to a commitment in 1985 by all Danube countries, including Austria and Germany, to monitor the river's water quality (Chapter 21). A number of bi-lateral agreements resolved river boundary questions or issues related to damage from water resource projects. By the end of the 1980s that citizen's movements began to develop, invigorating or replacing the government-funded conservation groups which did little to achieve their goals. A new green movement became politicized toward the end of the decade and helped to turn out communist governments.

7.4.1. MANAGING NAVIGATION ON THE DANUBE

The Congress of Paris convened in 1856 to negotiate peace after the Crimean War placed the lower Danube under an international commission. The Treaty of Paris signed by Austria, France, Great Britain, Prussia, Russia, Sardinia, and the Ottoman Empire at Galati, Romania in 1856, established the Danube Navigation Commission. Its authority was strengthened after World War I, following negotiations in 1921 and 1923 among all the riparian countries as well as Great Britain, Italy, Belgium and Greece; it was officially renamed the European Commission for the Danube. The Commission had its own flag, could levy taxes and provide diplomatic immunity to its members. It controlled navigation from Ulm to the Black Sea and kept navigational equipment in repair.

Disruption occasioned by World War II was followed by the development and ratification of another agreement, the Convention Regarding the regime of the Navigation on the Danube, signed in Belgrade in August 1948. While the US, Great Britain and France participated in the initial discussions, the east European states and the USSR agreed that only Danubian countries were to become Commission members. Austria became a member in 1960; the Federal Republic of Germany never joined but does participate in meetings as an observer.

The agreement affirmed that navigation on the Danube would be free for the nationals and vessels of all states and that port and navigation charges would be equally applied. A Danube Commission was established, with one representative from each Danubian State (with the exceptions noted above). The commission supervises the implementation of provisions, and the preparation of a plan for the works needed to maintain navigation equipment. Two special river administrations were established for operations on the lower Danube. The Commission is in charge of traffic regulations, pilot service, river inspection, and hydrometeorological services, provides statistics on navigation and trade, and publishes reference works, charts, and atlases.

7.4.2. POLLUTION MONITORING: THE BUCHAREST DECLARATION

Increasing evidence of pollution in shared tributaries and the alarming deterioration of the Black Sea in the 1980s directed attention to control of pollution on the Danube. At a meeting in Bucharest in 1985 the Danube countries, including Germany and Austria, signed a declaration agreeing to work toward reduced pollution in the river and to maintain a long-term monitoring program at eleven transboundary river sites countries. The monitoring, which recorded the rate of flow, temperature, and the levels of over 30 parameters, began in January of 1988. The Romanian Institute of Environmental Engineering Research, a government organization, became the focal point for the recording and sharing of data.

Despite the promise of corrective action, the declaration resulted in a minimum of analysis of the data collected and no agreement on treatment activities. The data are not shared among countries for almost a year after their collection precluding immediate response to increasingly serious pollution. Sampling methods have never been completely calibrated so that the data cannot be said to be compatible.

A multilateral monitoring agreement was reached by the countries in the Tisza River Basin in May of 1986. Representatives of Hungary, Czechoslovakia, Yugoslavia, Romania, and the USSR agreed to monitor conditions on the river, a major Danube tributary, using compatible sampling and analytical methods. The agreement stipulated that they would share data periodically and keep each other informed about accidental spills and other emergencies. The convention did not enter into force until 1990.

7.4.3. GREEN NGOs

During the period of Soviet domination, each of the Danube socialist countries had government sponsored pro-environment groups. In Bulgaria, for example, the "National Committee for the Preservation of the Environment," participated in and sponsored international conferences on environmental management and peace.

At an "Eco-Forum for Peace," I attended, which was sponsored by the National Committee and held in August 1986 in Varna, Bulgaria, over seventy papers explored the possibilities of international cooperation on environmental clean-up. Only one speaker, a Russian scientist, mentioned the Chernobyl disaster which had happened the month before. His principal message was that it must never happen again. The Committee organized a Danube cruise in 1988 with the invited participation of west European river experts. The goals of these activities were more to support so-called peace initiatives with the west and less to strengthen national and international environmental management policies.

Environmental groups critical of government environmental policies increased in number and strength in the 1980s. Air pollution from industrial plants, untreated wastewater, and massive waste dumps profoundly increased distrust of government environmental policies. In Slovenia, Romania, Hungary, and Bulgaria such groups became increasingly political (Jancar-Webster, 1991, pp. 192-219).

7.5. A Conflict Resolution Failure: Gabčikovo-Nagymaros

The challenge to communist rule posed by environmentalists gained strength in Hungary and to a much lesser extent in Czechoslovakia over reaction to the construction of the Gabčikovo-Nagymaros water resource project. Conceived by government agencies in Czechoslovakia and Hungary in the 1950s, the project appeared to have twin goals: to provide an independent source of energy for the two countries and to demonstrate the efficacy of socialist policies. No environmental impact assessment was made, although engineers from the two countries reviewed numerous designs for the intricate and massive project.

7.5.1. ORIGINAL PLAN

As originally planned and designed Gabčikovo-Nagymaros included three dams and two hydroelectric power plants. The Danube was to be dammed at Dunakilti on the Hungarian side of the river to flood an area of 60 sq. kms. [24 sq. mis.], ending at Bratislava. The water level of the reservoir would then rise 6.5 meters above ground and be routed downstream from the dam to a hydroelectric plant (capacity 720 MW) located at the second dam and lock at Gabčikovo. The river was to be diverted by a 17.5 kilometer parallel bypass. From this height, the water would have enough momentum to turn the eight turbines at the giant Gabcikovo facility. The water course was to be routed back to join its old bed.

One hundred kilometers downstream, the Danube bend creates a steep valley at Nagymaros where there was to be a second hydroelectric plant at the third dam (158 MW). This dam was to compensate for the twice-a-day release of water from the Dunakilti reservoir to the turbines at Gabcikovo. An agreement between the two countries specified that they would share on an equal basis the costs of construction and the anticipated electricity output of 3,775 KW hours per year.

7.5.2. AGREEMENT TO BUILD

An agreement to build the project was reached in September 1977 between Janos Kadar, the prime minister of Hungary, and Gustav Husak, president of Czechoslovakia. Ground was broken by a Czechoslovakian construction team early in 1978. Opposition to the dam because of the environmental damage it would bring to a significant part of the Danube, including the historic Danube Bend region and wetlands which protected Hungary's ground water supply, grew after extensive debates among Hungarian scientists. The Hungarian Academy of Sciences suggested in 1983 that construction should cease pending an environmental impact statement. More than 10,000 signatures opposing the dam were collected and a petition addressed to parliament.

Opposition and a Court Case

Demonstrations continued in Budapest and began to focus on the completion of the Hungarian end of the project while work continued on the Czechoslovakian side, without open opposition in that country until the late 1980s. Work was suspended on the Hungarian side and the parliament voted to withdraw from the project in the fall of 1989. The Hungarian government then abrogated a 1977 agreement to jointly build and operate the project, responding again to large public demonstrations. The Slovak Republic (SR), which experienced belated and not as vigorous public opposition, began in 1992 to divert water into the lengthy canal despite Hungarian opposition. Each country has appealed the action of the other to the International Court of Justice in the Hague. A decision was expected in 1997. The environmental effects of the project have received world wide attention (Fitzmaurice, 1996).

No Environmental Impact Assessment

The project is a monument to inadequate river basin planning. Its costs, monetary and environmental, have never been clearly acknowledged by its builders (an SR government company) and its funders (Austrian banks). Its benefits have never been itemized and their value supported. It appears neither to serve the goals of resource protection nor of increased energy supply. The project was supposed to provide cheap energy for the two countries, unimpeded navigation, and flood protection. In reality it would have supplied an estimated 2 to 3% of both countries' electric power needs and was not cheap enough to warrant the costs of construction (eventually estimated at well over a billion dollars). Navigation and flood problems had already been resolved.

As the canal was filled in 1992, a 200 sq. mi. area in Hungary known as Szigetkoz or "inland passage," was deprived of a resource which had nourished one of Europe's last wetlands. Only about one tenth to one fifth of the customary water volume has been allowed to flow into the Danube in northwestern Hungary. Some smaller branches of the river have dried out completely, and the groundwater level has dropped by as much as a meter in some places, and by two and a half in others. Recent literature, primarily from Slovakian sources, argues that there has been little damage to the environment from the operation of the dam and other aspects of the project. The continued dispute emphasizes the need for basinwide management of Danube resources, and demonstrates the lack of wisdom in development which ignores it.

7.6. Danube Evaluation, 1989

Agreement on common forum to solve problems and share data

- Agreement to negotiate basin conflicts was limited by the cold war although West Germany and Austria did accept, along with the East European countries, the navigation authority of the Danube Commission and participated in water quality monitoring under the Bucharest Declaration.
- Subjects for discussion and negotiation between western and eastern countries and sharing of data among them, however, was severely limited in quantity and content. The forum provided by the Soviet instituted CMEA was ineffective.

Binding convention for all countries

- A basin-wide convention to support long-term assurance of an equitable use of catchment area resources, delegating granting some authority to an international agency was not reached; the Danube Commission, remained primarily navigational in scope. Nor was there a decision to create an international commission with authority to regulate in these areas.

Basin-wide development plan

- A basinwide economic development plan which combined optimal use of natural resources in the interest of all concerned with adequate protection was not attempted during the cold war for a number of reasons.
- The third party attempts of the Soviet Union to coordinate economic and environmental coordination were no more successful than the belated and ineffective requirements of international donors.

Development and sharing of data

- Sharing of hydrologic and trade data took place to facilitate navigation under the guidance of the Danube Commission. Extensive monitoring networks were used for this purpose.
- Water quality monitoring under the Bucharest Declaration did not begin until the late 1980s, was confined to eleven transboundary river sites, seriously limited in its compatibility, and belatedly exchanged without any attempt to use it for management purposes.

Development and donor sensitivity to environmental impacts

- Water resource projects throughout the basin were not subjected to assessments which considered their impact on environmental conditions in the catchment area, including a significant number constructed in Austria and Germany.
- Investments by the Soviet Union in the East European countries were made, for the most part, without reference to environmental safeguards. Late in the 1980s international donors included such considerations in loan agreements.

PART THREE: A New Day for the Danube, 1990-1992

CHAPTER 8

Joint Danube Programs: 1990-1992

8.1. Western Concern

Even before the lessons of the Chernobyl explosion in 1986, international concern about the danger of air and water pollution in central and eastern Europe had steadily increased. Aid for environmental clean-up became a part of the western post-Cold War agenda. The EU Phare program funded waste treatment programs and made improved environmental policies a membership criterion. UN agencies sponsored a global approach for resource protection in which all Danube countries participated. And, by the end of 1990 plans were set for the first phase of an international environmental program to restore and help protect Danube resources.

8.1.1. INITIAL WESTERN AID AVAILABLE

Initial funding for eastern Europe after the overthrow of Communist governments as available from a number of international sources for humanitarian assistance, financial help for the stabilization of economies, agricultural and industrial reforms, and rebuilding of infrastructures. The euphoria following the end of communism generated an assumption that "a three to five-year burst of external assistance--balance of payments and structural adjustment support from the international financial institutions, bilateral technical assistance programs coordinated through the G-24 and facilitation of foreign trade and investment--would propel these countries toward integration into the global economy (Hutchings, 1994, p. 184)."

The euphoria, of course, dissipated over time. Coordination among the programs was not always discernible and donor levels of concern for the environment not always clear but, as it turned out, there was help available for a basinwide program to restore and protect damaged Danube resources.

Group of Twenty-Four

In 1989 the Group of Seven (G-7) leaders established the Group of 24, members of the OECD, EU members and other industrialized democracies, with the EC as its chair. It was to coordinate western aid to Eastern Europe. Working units directed aid for improving food production among other urgent programs, initially to Hungary, and later to other Danube countries. From 1989 to 1992 the G-7 had committed nearly $35 billion to the Central and East European (CEE) nations. Of this about 42% was grant aid and the remainder loans. By May 1990 G-24 had committed $8.5 billion in CEE aid to Hungary and Poland. The EU Phare program committed a total of ECU 2.29 billion from 1990 to 1992 to the Danube countries.

World Bank

The World Bank had made loans available to its east European member countries during the post-World War II period. Czechoslovakia was an original member but left in 1954, rejoining in 1990. Yugoslavia, also a founding member, had received 90 loans of over $6 billion and had about $2 billion remaining in its account at the start of the transition; it ceased to be a member in 1992. Romania had been a member since 1972 but lending had been discontinued in 1982. Hungary, a member since 1982, had an outstanding credit of almost $800 million in 1989 (Aziz, 1994, p. 192-193). The Bank was ready to provide technical assistance to member countries. ." to set up projects that might be suitable for loans. Eventually all of the eligible Danube countries became members or restored their membership status. From 1990 to 1992 the World Bank disbursed $2.4 billion in nonconcessional loans to the five EE countries.

EBRD

The European Bank for Reconstruction and Development (EBRD) was established in May 1990 by 40 countries, the European Union, and the European Investment Bank. The EBRD had a special mandate to provide assistance to the former communist country. It also supported environmentally sound and sustainable development, democracy, and nascent market economies. Its initial capital ($12 billion ECU) was considerably smaller than that of the IMF.

US AID

United States funding came under the Support for Eastern Europe Democracy (SEED) program which obligated $1.5 billion in aid from 1989 to 1992; of this amount almost half went to Poland (US Department of State, 1994, table on p. 278). SEED had three primary objectives: to develop market economies and a strong private sector; develop and strengthen democratic institutions to sustain democracy, and improve the basic quality of life in priority geographical areas.

8.2. Support for Environmental Programs

East and west had joined together in regional environmental agreements during the years after World War II. The United Nations Conference on the Human Environment, held in Stockholm in 1972, was one of a series of initiatives which encouraged the sharing of information and coordinated environmental action in Europe and with other parts of the world. Other agreements and international conventions signed by both eastern and western governments promised greater transparency in national environmental protection policies in Europe and globally. In reality most of the commitments required compatible intra-national monitoring of pollution levels if they were to be effective. Enforcement of standards was limited, particularly in the east European countries.

8.2.1. EC SUPPORT

A May 1990 EC Council directive established the European Environment Agency (EEA) and the European environment information and observation network (EC, 11/5/90). The EEA goal was to provide the Community and its member states with "objective, reliable and comparable information at European level enabling them to take the requisite measures to protect the environment, to assess the results of such measures and to ensure that the public is properly informed about the state of the environment, to that end, the necessary technical and scientific support (EC, 11/5/90)." The EEA opened its doors in Copenhagen in 1993 to carry out the broad EU mandate.

Dobris Assessment

Ministers from east and west participated in a meeting in Dobris Castle, near Prague, in June 1991, to review the proposed development of an environmental program for Europe. It requested a report on the state of the environment as an initial step. The Dobris Assessment and its statistical compendium, published in 1995 (Stanners, 1995 and Europe's Environment: Statistical Compendium for the Dobris Assessment, 1995), are meant to support a long-term program as well as the more immediate Environment Action Programme (EAP) for Central and Eastern Europe. The Programme was endorsed at a ministerial conference in Lucerne, Switzerland, April 1993.

A wide range of issues is covered in the Dobris Assessment which was designed to identify priorities for repairing environmental damage and to support policies to prevent future problems. It describes and analyzes environmental trends throughout Europe, emphasizing the interplay between environment and pressures on it. The compendium has over 400 pages of statistical tables, covering a wide range of topics related to air, water, and noise pollution, population trends, and economic activities. Altogether it is an achievement which, with periodic updates, can be a guide for policy throughout western and eastern Europe in the decades ahead.

Environmental Action Program

The Environment Action Program for Central and Eastern Europe calls for national programs to be prepared by each CEE country. These are to follow a common approach to support a framework of optimal environmental policy, institutional strengthening and investments (Regional Environmental Center, 1995). The National Environmental Action Plans requested by the Program complemented a number of early transition fact-finding programs. In 1990 and 1991 teams made up of World Bank, US AID and Environmental Protection Agency experts, traveled to CEE countries and reported on the effectiveness of environmental policies. National reports were provided to the UN Conference on Environment and Development held in Brazil in June 1992.

Statistical offices in several countries published yearbooks of environmental statistics.

Global Environment Facility

The establishment of the Global Environmental Facility (GEF) in 1991 was a response to the need for closer donor attention to the environmental impact of international grant aid and investments. A pilot program was launched by three implementing agencies; the World Bank, and the UN Development and Environment Programs, to assist developing countries and those with economies in transition in the protection of biodiversity and international waters, reduction of global warming, and projects related to the depletion of the ozone layer.

The initiative began in July 1991 with funding of $1.2 billion to be allocated among the four objectives. The funds were to be added to regular development assistance and provided as untied grants or loans on highly concessional terms to countries with a per capita GNP under $4,000. An advisory panel of experts guided the funding which was to be limited to financing the incremental costs required to achieve global benefits.

8.2.2. DANUBE RIVER BASIN TRANSITION NEEDS

Most of the east European countries and two of the former Soviet republics in the Danube Basin were eligible for western assistance and for help with environment programs. Transboundary air and water pollution was a basinwide problem subject to regional solutions. The Danube Basin agreements then in effect clearly could not deal with mounting transboundary contamination in tributaries, the main stream, and the Black Sea. Assistance from the west to the newly democratic countries of the Basin provided a unique opportunity to bring improved management to the Basin. Coordinated management of international river basin resources, applied to the Danube, could advance the restoration and protection of ecological resources. Such institutional strengthening could help as well as benefit from political and economic reform.

Management Format

The basin lacked an effective management structure to deal with mounting water quality and other environmental problems. Its countries needed to engage in a process familiar to other international river basins. They had to initiate many of the steps described in Chapter 2. Specifically, they had to:

- strengthen substantially the relatively soft compact to monitor at eleven transboundary sites on the mainstream of the river;
- find ways to develop accurate data about intranational environmental quality and share it freely with each other;
- commit to a basin-wide convention which would go beyond the regulation of and navigation to protect and improve the management
- of basin ecological resources, and agree on a plan to prioritize

clean-up needs and support the sustainable development of basin resources.

By the end of 1991 Danube countries had agreed on specific ways to reach most of these four objectives. They had guidance and funding from the EC which had joined the International Commission for the Protection of the Rhine, and from the GEF, the EBRD, and US AID. During 1991, representatives of the Danube countries and the international agencies met twice to advance joint action in the basin: once in February in Budapest to review and discuss proposals for a formal Basin convention and again in September to establish the framework for joint action programs to guide investment in natural resource protection, monitor water quality and share information about it, and encourage the strengthening of institutions to support these and other basin management goals.

8.3. An all-Danube Convention

The time was ripe by 1990 and 1991 for Danube countries to begin negotiations on an agreement on an international convention which would advance the protection of Danube environmental quality. International meetings and discussions had paved the way. In 1989 the Conference on Security and Cooperation in Europe (CSCE) which had been founded in 1973 to help promote detente between the west and the east, met in Sofia to discuss the possibility of such a convention. Participants from the Danube countries recommended the development of a comprehensive framework agreement to resolve pollution and flow issues in transboundary streams. As a start, Bulgaria and German agreed to pursue an "east-west" Danube waste management initiative.

Other international activities were evidence of the willingness of east and west to commit to mutual environmental management goals. In 1991 the UN Economic Commission for Europe (UNECE) formulated a model framework convention for transboundary rivers and international lakes. Submitted to a CSCE meeting in Helsinki in 1992 it served as a model for the convention eventually accepted by the Danube countries.

8.3.1. HUNGARY PROVIDES A FORUM

In February 1991 the Hungarian Ministry of Environment provided a forum in Budapest for intensive consideration of two proposed convention concepts. All of the Danube countries were represented. At that time they were Austria, Bulgaria, Czechoslovakia, Germany, Hungary, Romania, the Soviet Union (for Ukraine), and Yugoslavia. International agencies represented included the EC, the UNDP, UNECE, UNESCO, World Health Organization, the Rhine Commission, and the new EBRD.

Two conventions were presented in outline form: one by the Hungarian delegation and one by the Austrian. Each country continued to pursue separate conventions in the years following the meeting. Early on the Austrian version which focussed on water quality issues received wider support although discussions continued on the Hungarian proposal.

Hungarian Danube Convention Concept

At the February meeting the Hungarian delegation headed by Dr. Janos Tardy, of the

Ministry of Environment, called for an ecological rather than water quality approach. The proposal cited for support the CSCE Sofia meeting as well as the all-European Convention proposed by the UNECE. A convention objective should be "to achieve by specific measures that no interference, development or improvement of the actual conditions shall be undertaken in the Danube region, unless the aspects of ecology are taken into consideration, further to declare that protection and maintenance of the wide diversity in habitats and wildlife of the Danube is in the long-term interest of the users of the area (Hungarian delegation, 1991)."

The proposed convention emphasized cooperation to ensure ecological protection from all forms of pollution, in all media. Its concern reflected priority problems identified by Hungary, which depends on sources outside the country for 95% of its drinking water. Threats to its water resources from the large dam at Gabcikovo were a paramount issue in the country at the time. The Hungarian proposal required an inventory of pollution problems and an ecological evaluation of agreements and crossboundary projects with special priority given to "areas of particular public concern" (e.g. Bos [Gabcikovo], the Danube Delta, etc.). Hungary offered a geographical base for a temporary secretariat to prepare text, set up working groups on the most severe ecological problems, and a foundation for financial support.

The Austrian Version

The Austrian representative, Wilhelm Kittinger, from the Ministry for Agriculture and Forestry, proposed an agreement which emphasized protection of water quality, called the "Convention on the Protection and Use of River Danube." The objective emphasized the need to strengthen national and international measures to improve or maintain the quality of the river, its main tributaries and its estuary; to ensure usability of water resources, especially for drinking and irrigation. It also suggested the need to "prevent, control and reduce the discharge of hazardous substances and of nutrients from point and non-point sources" and "to develop and apply water quality criteria, individually adjusted to specific reaches of the Danube river and their relevant water resources as ecosystems."

Suggested "operative elements" called for a stipulation that municipalities should be equipped with sewage treatment plants at the "minimum stage of biological purification and the adoption of discharge licenses prior to discharge of any pollutants." Terms of reference were suggested for the establishment of a permanent Danube Protection Commission with a permanent secretariat. The Austrian version, with the short name of Danube River Protection Convention, for which serious negotiation began in 1992, was signed by nine Danube countries in 1994 and had been ratified by six by the end of 1996. It is expected to enter into force in 1997 (Chapter 21).

8.3.2. AGREEING ON A DANUBE ACTION PROGRAM

Simultaneously with the negotiation of a new Danube-wide convention, staff of the World Bank, UNDP, and the EC were getting together a program for more effective management of Danube River Basin resources which would support convention goals.

Whether they favored the Austrian or the Hungarian version, participants in the February Budapest meeting also wanted a funded action program for the Basin. There was general agreement that such a program need not wait for a convention to be negotiated. When the latter went into force the program would become part of the operating mechanism for the proposed International Commission for the Protection of the Danube River. Experts from the major funding sources, the EC, GEF agencies, and the EBRD, made plans for a "founding meeting" which would establish a joint action program to be operated by international organizations and Danube country representatives.

UNDP Support

The UNDP had been interested in coordinating help to restore Danube Basin resources and provide long term protection for them since the 1980s. In 1987 it sponsored a meeting of Danube countries which endorsed a River Danube Water Quality Protection project. Funding was limited; the project was expected to receive no more than $17,000. Another UN group, the World Health Organization (WHO), continued the project by financing a number of consultancy missions to the region. Bulgaria, Romania and Yugoslavia signed a revised project document but the proposed coordinating secretariat created political problems which kept UNDP from proceeding. The agency quit the project when it appeared that a new all-basin convention was to receive support from riparian countries.

By 1991 the UNDP had decided to "concentrate on preparing the ground for future regional cooperation on the Danube by organizing a survey and a feasibility study for the management of water resources in the Danube (UNDP, 1991)." As part of the Global Environmental Facility it had increased leverage to pursue its goal.

8.3.3. THE SOFIA MEETING

Format

The gathering in Sofia in September 1991 was a unique meeting from more than one point of view. At an early point in the transition period it brought donors and donees together to agree on a regional approach to urgent environmental issues. It was one of the first times, in any part of the world, that all river basin countries were represented at a meeting with donors in advance of major decisions about basin investments. And it certainly was the first time that the majority of the countries involved in planning joint environmental policies for an international river basin were also implementing drastic changes in ideology and political and economic processes.

Discussions about the September Sofia meeting format were held at the February 1991 meeting in Budapest. In particular the representative of the UNDP, Kees Wijnen (pronounced Case Wynan) discussed these possibilities with some participants, including myself, about the possibility of holding a technical meeting of country representatives in tandem with a plenary session of countries and donors. My Fulbright grant had brought me to the Bulgarian Academy of Sciences in Sofia in September

1990 to develop a water quality information system for the basin. I had become familiar with environmental legislation in the countries and their ministries of environment. The strategy was to hold two meetings. At one donors and country representatives would hold an initial plenary session to discuss basin environmental problems. Then technical specialists from each country would meet in working sessions to identify major needs for a basinwide environmental program. Ian Dogterom, director of the International Center for Water Studies (ICWS), Amsterdam, and I worked on the technical sessions. The Bulgarian Academy of Sciences hosted both meetings.

Participants
Representatives from the sponsoring international funding organizations, all of the Danube countries, and international non-governmental organizations (NGOs) especially concerned with the environment, assembled in Sofia in late September. Plenary sessions were to discuss a proposed Environmental Program for the Danube River Basin (Danube Program) while separate technical expert sessions would review the need for and the resources available for such a program. Ideally, two representatives were to come from each country: one an official representing a ministry with environmental policy responsibilities, and an expert from a research institute.

Meeting Agendas
The expert meeting was divided into three concurrent work sessions. Session I adopted recommendations for the updating of a basic information system package for the river basin. The second work session reviewed legal water quality standards and their enforcement and made recommendations for their harmonization among countries. The third featured analyses of recent water quality monitoring data and laboratory methodology, to stress the need for a comprehensive water quality data monitoring system throughout the basin. The work sessions reported findings to the plenary session at the second, or program coordination, phase of the meeting.

8.3.4. EXPERT RECOMMENDATIONS

Data Management System
A data management system was proposed which was to serve as a problem solving tool, as well as providing quick access to reliable data and information about the environment of the basin. The system's databases were to include a bibliography, directory of institutions and experts, factual data, standards and regulations, among others. Public access was to be provided to data and information derived from the system which would be based on data collection and dissemination by international networks. The system, it was agreed, would be valueless if it was not used for the analysis and evaluation of basin environmental policies.

Water quality legislation
Overall goals for the Program included the establishment of effective environmental

standards for the catchment area, adoption of policies which would treat the river basin as a significant natural resource, and the enforcement of compatible standards in all the countries, taking into account overall economic development needs as well as those of the individual countries. At short term, the Program was to prepare relevant databases, including those of existing environmental standards in the riparian countries, and relevant features of national legislation. An immediate start to this aspect of Program activities was urged.

International Water Quality Monitoring Program
A number of objectives were identified for an international water quality monitoring program:

- identification of major sources of pollution;
- trend detection in concentrations and loads of priority pollutants;
- assessment of water quality pollution and its effect on human health particularly with respect to drinking water, and
- the production of periodic water quality status reports.

To achieve these objectives it was recommended that the existing Bucharest Declaration monitoring programme be evaluated and subsequently improved and extended where necessary.

8.4. Commitments for the Future

On the final days of the meeting, the 26th and 27th of September, the international agency representatives responded to the recommendations of the experts session with commitments to fund and support in other ways the activities of the proposed Danube Program. A three year action plan for environmental management would constitute the first phase of this international cooperation and would:

- develop a series of focal points to support program management;
  prepare pre-investment studies on a national basis and complementary studies on special topics to support development of a Strategic Action Plan;
- strengthening and "human resources" development activities;
- provide for a series of fact finding, institutional work to harmonize legislation;
- prepare an emission inventory;
- strengthen and extend monitoring programs and. laboratory facilities;
  strengthen data management. inventory biological resources, and establish networks of specialized institutions, experts and non-governmental organizations concerned with environmental management in the Danube River Basin.

The three year plan was to set these activities in motion beginning in early 1992. A number of specific agenda items were suggested.

1) The program would support analyses and activities for integration of nature conservation and other measures which recognize the interrelationships between air,

soil and water into an environmental management framework for the Danube River Basin.

2) To ensure effective and widespread support for the program, the public should be appropriately informed of the objectives of the program and involved in a participatory manner. Nongovernmental organizations would participate actively in a variety of aspects of the program.

3) A second phase of the program would include large scale investment activities, further institutional strengthening and continued human resources development.

The second phase would implement the recommendations of a "Strategic Action Plan," which identified priority rehabilitation needs and began pre-investment studies for projects during the first phase. Priority actions were to include support for the Bucharest Declaration, the conclusions of the Dobris Environment Ministers Meeting, and the future Danube Convention and Agreement on Environment.

An Environmental Programme for the Danube River Basin jointly supported by the riparian countries and the international community was to be established immediately on a cooperative basis, as was an interim Task Force, comprised of riparian government representatives and other relevant parties. Funding and probable amounts of contributions were discussed at the meeting. Final decisions on funding are included in the next chapter.

CHAPTER 9

The Danube Program, Phase I

9.1. Initial Plans

When the first meeting of the Danube Program Task Force was held in February 1992 a small staff had provided a lengthy work plan complete with budget information and detailed outlines of proposed activities for the next three years (EPDRB, 1992). There had been some changes in the membership of the Task Force. Yugoslavia was still a member but would not be for long in the face of the trade embargo and other sanctions which the UN and EC were to declare in April 1992. Croatia and Slovenia, which had proclaimed their independence in June 1991, joined the task force. Eventually the Soviet Union no longer represented Ukraine which become an independent republic in 1991 and thereafter represented itself.

9.1.1. WORK TASKS

Funding had been committed and a detailed work plan modeled on the conclusions of the Sofia meeting. There were serious problems facing the basin. The violent dissolution of the Yugoslavian Federation and its devastating effect on human lives and the regional economy cast a shadow over all efforts toward international cooperation in the Basin. Each of the formerly socialist countries had formidable obstacles to overcome before economies and governments could be considered stable. The new Danube Program Coordinating Unit (PCU) would be asking governments to develop and tap resources for another set of reforms: revising inadequate environmental policies. The political and economic processes of some countries were closer to those of western Europe than others. None had adequate environmental monitoring and data technology, waste treatment facilities, or laws compatible with those in EU countries.

There were other problems. The PCU staff was small. Its organization scheme called for three specialists and "adequate administrative and secretarial staff." The team leader and other staff were on contract and would be responsible to respective funders. While the work plan identified the length of time tasks were to take and the funding which was allocated for them, basic decisions about procedure, including the selection of personnel, were dependent on decisions made by the funders.

9.1.2. PROGRAM SUPPORT

The eastern European countries were eligible for GEF, EC Phare Program, and EBRD aid. The indebtedness of most of the country precluded borrowing for environmental facility investments. The "work" part of the work plan (EPDRB, 1992) fell on the shoulders of ministry of environment staff members in each participating country who were familiar with the subject matter of the plan and had a command of English, the language used by the Program. Their presence was required at innumerable meetings. At home they had their usual work to accomplish plus whatever assignment had come their way from the PCU and the various sub-groups it formed. "Cooperation" in the Program initially implied that participant countries would fund staff time. Those working on EPDRB activities had staff and budget problems plus the uncertainty resulting from changes in governments and fluctuating ministerial support for their activities.

Responding to the Sofia Meeting

The Work Plan presented for Task Force adoption in February committed government representatives and international funders to a detailed set of activities. The plan followed the goals and objectives supported at the September Sofia meeting. It divided activities into two phases: an initial, which would accomplish priority objectives identified in Sofia and a second, investment phase, when waste treatment facilities and other infrastructure needs related to environmental protection would be funded and constructed.

The long term objective of achieving the sustainable use and development of the Danube Basin's natural resources will be addressed within the framework of the convention and the agreement currently being developed by the riparians. In support of this, the objective of the Regional Environmental Programme is to establish the operational basis for strategic and integrated management of the Danube River Basin environment while focusing initially on priority environmental issues (EPDRB, 1992).

Key Principles

Some key principles applied. The primary objective was to protect and enhance environmental values while contributing to economic welfare and safeguarding public health. The Danube Basin was to be regarded as an ecological system. The program was to encourage the participation of the Danube countries and reflect their priorities while emphasizing consensus building, information sharing, and other forms of joint decision making.

The Work Plan made it clear that all of the conclusions of the Sofia meeting were on the agenda: to support institutional and technical capacities in the countries with emphasis on environmental monitoring and data management, harmonized environmental standards and procedures, and a free exchange of data and technical information. In the initial phase of Program investment-related activities the development of a Strategic Action Plan would be emphasized, identifying the areas of acute environmental concern.

9.2. Benchmarks for Success

Detailed short and longer term activities, accompanied by time schedules and lists of products, suggested benchmarks by which Program achievements might be evaluated. Responsibilities and activities are summarized below.

9.2.1. STRUCTURE

The three-person professional staff of the PCU was to maintain a network of focal points in each participating Danube country; facilitate G-24 involvement; monitor all Program activities; establish an information sharing mechanism and periodically report to the Task Force on Program activities.

Short term actions were to identify high priority pre-investment studies and establish an accident emergency warning system.

Longer term activities, the Strategic Action Plan and investment program, would include an ambitious number of assorted assignments for donors and countries: national reviews of environmental conditions by each basin country accomplished in accordance with detailed specifications; pre-investment studies of high priority sites which would be assembled for funding during the second or investment phase of the program; regional studies of problems related to the basin as a whole, e.g. non-point pollution sources, air pollution and ground-water management.

Institutional and technical support for environmental planning and management would be provided and would include the development of analytical tools for planning and management.

Networks of experts would be developed, research institutions themselves strengthened as well as non-governmental organizations concerned about the environment.

9.2.2. ACTION PLANS

High priority pre-investment studies were to be part of a screening process to determine which sites should have priority during the second or investment phase. A small number of pilot demonstration projects would include not only civil works but the industrial and policy reforms essential for effective project financing, implementation and operation; because of time limitations the pilot projects would probably be associated with existing facilities.

Strategic Action Plan and Investment Program

The Plan was intended to support a comprehensive multiyear environmental clean-up program for the region. To that end it needed an assessment of the "nature, extent and causes of environmental degradation in the Danube River Basin." An assessment was to include "priority areas for intervention" in policy areas as well as investment activities. Developing and accessing needed information and data was a difficult task from many viewpoints. Ideally, the countries would have developed compatible, state-of-the-art monitoring, and laboratory testing and data-sharing techniques first and then followed the process with identification of priority investment sites ("hot spots"). First-phase activities were all concurrent, however, and to an extent sacrificed accuracy to expedience.

Pre-investment project preparation identified in the short-term activity described above and in the national reviews were to be prepared by consultants with the direct participation of multilateral and bilateral financial institutions in the planning, supervision and review of these studies.

Regional studies were to develop innovative techniques for monitoring the cumulative effects of pollution and changes in sub-regional and regional environmental quality. One technique was detection of radiation accumulation through gamma ray spectroscopy, and another the measurement of pollutants accumulated in the tissue of freshwater mussels in the riverine environment. Other regional studies might include the impact and management of non-point source and air pollution on the basin; reviews of municipal administration and finance of water supply and wastewater and solid waste management, and environmental issues in industrial restructuring.

Institutional and Technical Support

Harmonization of legislation and standards was to be aided by an information system which would include existing national standards and relevant legislative texts; long term action in this area was to be undertaken within the context of the future convention.

Environmental Monitoring

Environmental monitoring, sampling, selection of parameters, and strengthening of laboratory methods, was to be accompanied by improved data management: there was a need for "a more strategic and comprehensive approach to data collection, processing and exchange" for the basin. Decision systems should be developed; a computerized infrastructure for efficient data input, use, and exchange; the results should be made available to the public. A sub-group of the Task Force was to be created for this purpose.

Institutional support also included the strengthening of non-governmental organizations. A network of environmental NGO would receive assistance to ensure public awareness and involvement in the decision making processes. A "structured series of applied research activities" was promised. Care would be taken to avoid duplication of effort and ensure targeted results toward high priority tasks.

Funding

According to the 1992 Danube Program Work Plan the estimated cost of Phase I, the initial three year action plan, was to be 48 MECU ($57.6 million) most of which (41 MECU or $49.2 million) would come from international sources. Commitments made at the time totaled about 34 MECU ($40.8 million); this amount never materialized. The program began with commitments of somewhat less and ended with an even smaller amount to pay for its activities.

Table 9.1 Funding of Danube Program Activities, 1992-1995*

Funding Source	Amount (MECU)	$ millions	%age	Activity
EC Phare*	2.01	2.412	15	EPDRB Coordination
	1.07	1.284	8	Accident and Emergency Warning System
	4.42	5.304	33	Applied Research
	1.21	1.452	9	Regional Studies
	4.56	5.472	34	Monitoring, Laboratory and Information System
	0.13	0.156	1	Public Participation
Total	13.4	16.08	100	
GEF	3.04	3.648	49	World Bank pre-investment studies
	1.05	1.26	17	EPDRB Coordination
	0.68	0.816	11	Environment Ministries
	0.31	0.372	5	Sectoral Ministries
	0.43	0.516	7	Municipal Authorities
	0.37	0.444	6	Data Management Sub Group
	0.31	0.372	5	Public Participation
TOTAL 2	6.2	7.44	100	

*EU Tacis in 1996 provided Moldova and Ukraine with MECU .3 for Danube Program activities

Source: EPDRB 1996; report states that other funding sources had not made data available.

* * * * * *

The basin had the start of a management program and the promise of more by the end of 1992. The countries had a negotiating forum, with modest funding from the west to help. Workshops and collegial activity at the technical level proceeded amicably among all the countries even with the greater support they had expected for help with Program activities, the cost of which they for the most part bore by themselves.

PART FOUR: Politics, Economics and the Environment

CHAPTER 10

Setting New Agendas in the Danube Basin

10.1. Decision Cycles

International river basin management is an uninterrupted cycle of decisions and activities, each dependent on the other and on a number of historic, cultural, economic and other factors. The replacement of authoritarian regimes with democratic institutions may or may not be a condition for improved management of river basin resources in other parts of the world. It was an absolute condition for the restoration and long-term protection of the environmental quality of the Danube Basin as its countries emerged from the twilight of the years under communism. Pursuit of goals set by the Soviet Union made agreement to harmonize environmental policies with western Europe impossible. The Cold War had divided the basin between east and west so that there was little room for an accommodation on environmental .

When basin-wide agreement became a possibility the planned steps toward improved management of transboundary pollution initiated by international agencies were specific and wide ranging. Modest funding of joint programs, however, made it clear from the start that cooperation depended on substantial changes in political processes, steady improvement in economies, and the ability of each of the new democratic governments to adopt and enforce western environmental standards.

10.1.1. CHARTING POLICY CHANGES

The rationale for optimal international river basin management (Chapter 2) rests on three major criteria. Countries must understand not only their own river-related resource problems but those of all their neighbors, and of the basin as an entity. Possible solutions must be developed and implemented by agreement among the basin countries. Five river basins in four continents were evaluated on the basis of their respective abilities to reach the optimal basin-wide contract. It was recognized that the evaluation served a limited purpose. Details about national decisions on domestic and foreign policies are needed to evaluate river basin management needs and to determine the level of support available for long-run commitment to effective agreements.

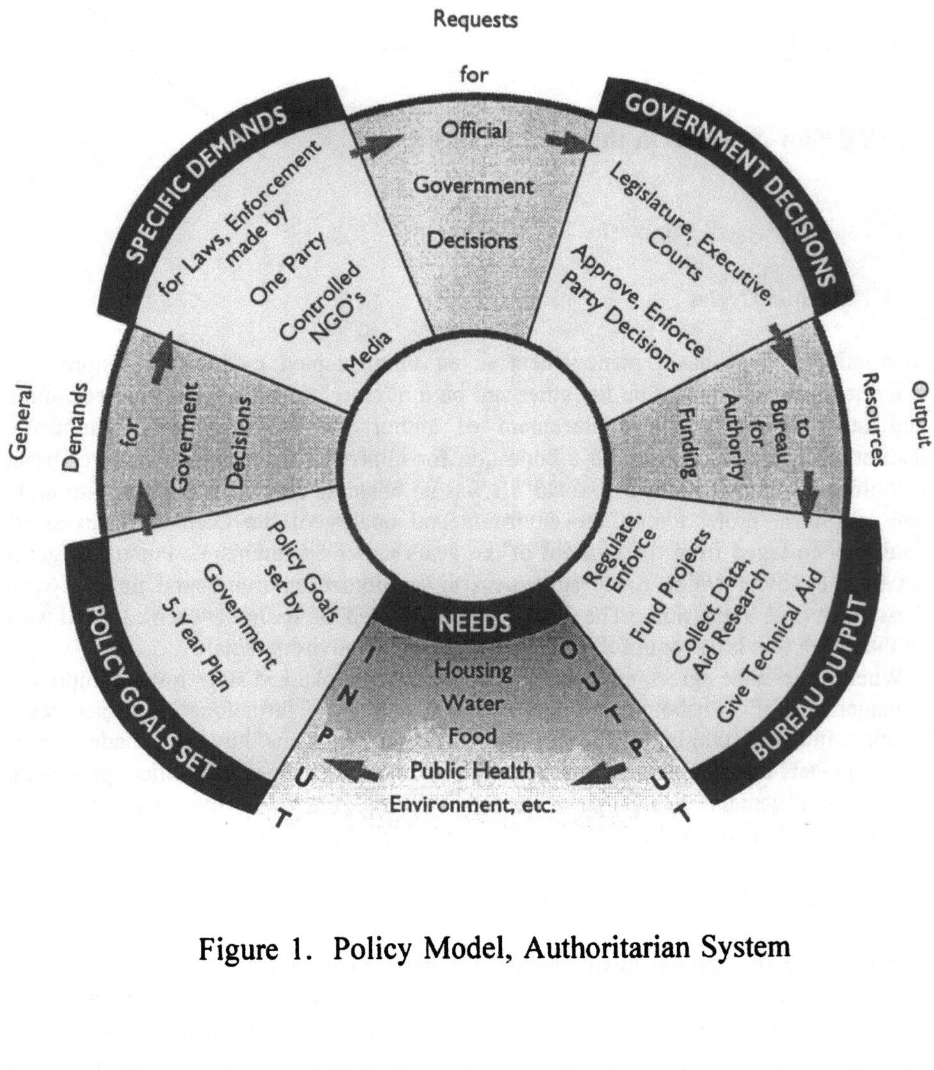

Figure 1. Policy Model, Authoritarian System

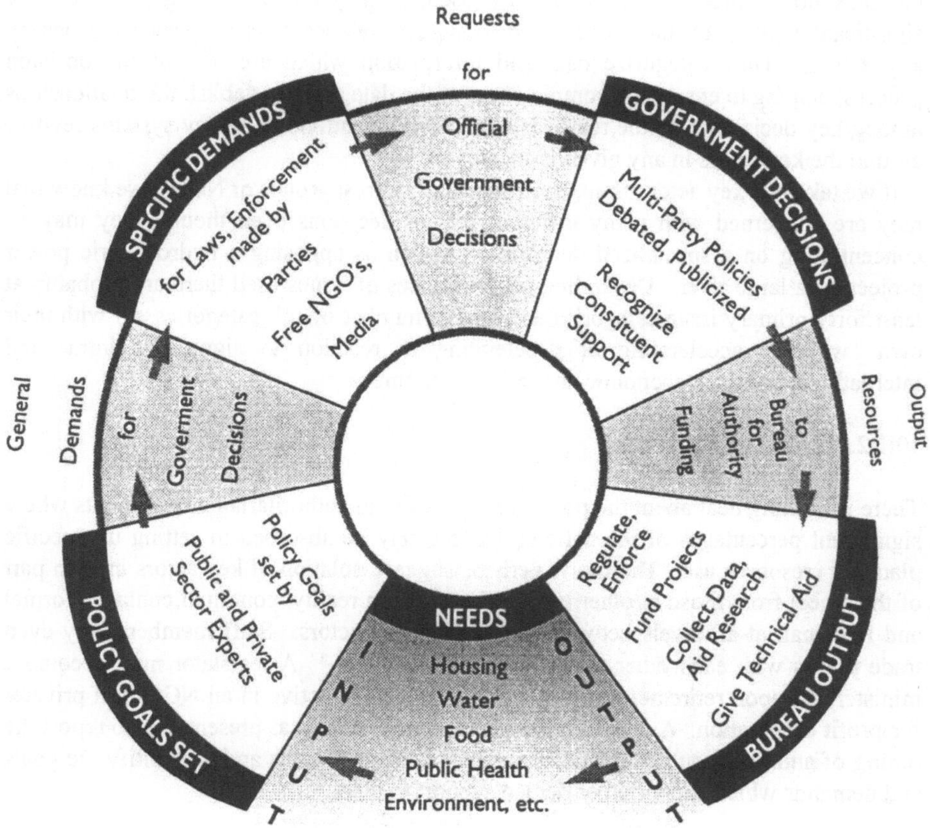

Figure 2. Policy Model, Democratic System

Government decisions follow the same general sequence whether they are authoritarian or democratic (Figures 1 and 2). The decision models over-simplify a complex policy process. They are best used, as they are in this study, to identify significant impacts on the process and to suggest relationships between decisions by key actors. They categorize data and information which are part of the decision process, helping to ensure the completeness of the data and to establish the relationships among key decisions. In the real world, a galaxy of individual policy systems revolve around the key actors in any given issue area.

If we take one key actor group, environmental interest groups or NGOs, we know that they are concerned with many different sets of decisions even though they may be concentrating on a specialized decision area such as opposing a hydroelectric power project on a large river. Depending on the indices of issues used there are probably at least forty primary issue categories and a large number of sub-categories, all with their own "wheels," accelerating or decelerating in reaction to significant intra- and international political, economic, and other phenomena.

10.1.2. REVOLVING DOORS

There is nothing neat about the policy process even in authoritarian governments where significant percentages of the national budget may be absorbed in setting up specific plans for resource use. The charts perhaps suggest isolation of key actors in each part of the wheel from those in other parts. There are, in reality, continual contacts, formal and informal, at all levels between and among key actors. Staff members may even trade places with each other through "revolving doors." A legislator may become a minister, or, upon retirement from government become active in an NGO or a private-for-profit corporation. Again, the importance of the schema as presented is to report the timing of and occasion for key decisions by government units and to identify the goals and demands which supported or denied support to them.

10.1.3. MEASURING DECISION IMPACT

Impacts on needs for government action can be matched with decisions. For example, transition watchers in the EC and elsewhere monitor the adoption and content of constitutions, the fairness of multi-part elections, and the like. With respect to water quality, pollution levels "before and after" government decisions to monitor and treat the sources of contamination may even provide statistical measurements of the success of the decisions (Figure 3). A previous application of the model to three large international river systems (Murphy and Sabadell, 1986) demonstrated the need for foreign and domestic policy goals to be mutually supportive of international commitments (Figure 4). A complex policy schema for a large basin would relate foreign and domestic policies of each country and show each in "harmony" and supporting an international "bureau" (commission) to implement mutually agreeable decisions.

In the case of the Danube the graphic display would show eleven out of thirteen two-wheel bi-cycles operating, for the moment in any case, in tandem, willing to provide resources (authority and funding) to an international commission.

The following chapters apply the decision system model to the east European and CIS countries presently in transition, adapting political, economic, and environmental policies to western models. A description of the format used for each follows.

10.2. Changing the Political Process

Figures 1 and 2 emphasize the magnitude of the political changes taking place in the former Communist Danube countries. The decision wheel depicts responses to issues which take place in all types of political systems. When needs reach a certain level governments must react. Optimally, previous decisions have assured enough resources to a bureau so that its response to a request for a product or service is automatic. There is no need for a decision to add to its resources. Access to drinking water is a universal need which obviously must have a high priority with governments. Increasing scarcity, growing pollution from agriculture and industry, and other factors can make it difficult for "water bureaus" to satisfy thirsty customers or drought stricken farmers. Some of the inadequacies caused by failure to treat or prevent pollution in Danube countries are described in Part 2. These and other issues as they affected decisions to participate in the Danube Convention will be discussed in the following country profiles.

10.2.1. EFFECTING CHANGE

Each of the parts of the decision wheel must undergo profound changes to complete a transformation to a democratic system. Goal setters had to be freed from the uni-party supervision of the past and find support elsewhere, perhaps partially from government budgets but primarily from the private sector. The civil society, as measured by multiple political parties, interest groups relating effectively to constituents, and a free and independent media, has visible problems in several countries. The governments themselves are also experiencing pressures. Clearly financial constraints are the occasion for hard bargaining for annual budget allocations.

Bureaus, which have been provided with budgets and authority, create outputs. These have been arbitrarily divided into four categories with overlapping functions. There may be more than one bureau and more than one ministry working on the same issues. There is a bibliography on the subject of "coordination" among water resource bureaus and agencies. For one interesting example see OECD, 1989, the result of a three-year study to improve integration of water policies among government agencies.

The first major political reform during the transition, the adoption of new constitutions by most of the countries, was intended to change the functions of each part of the process. New constitutions sanctioned rule by law and not by a single party; protection of the right of individuals and groups to disagree, and the obligation of governments to make the decision process itself and its products transparent for all.

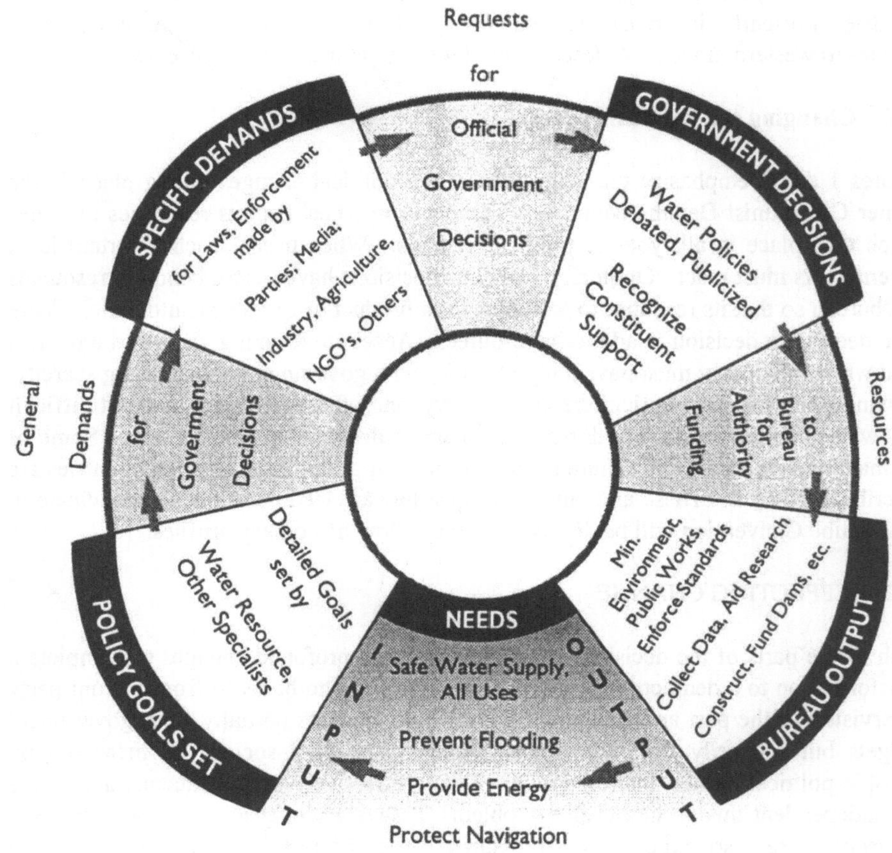

Figure 3. Water Resource Policy Model

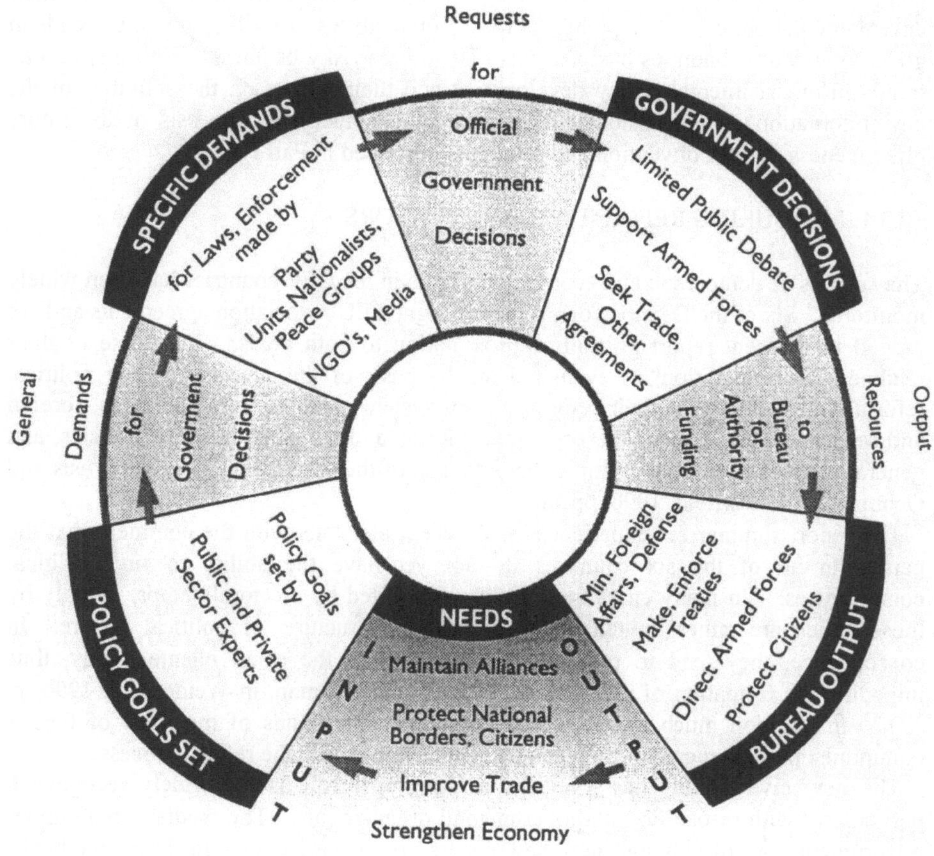

4. Foreign Policy Model

10.3. Profiles of Change

The following chapters summarize the political and economic changes in each country from 1990 to 1995 as a background for a more detailed review of their environmental policies. "Stopping the clock" on analyses of post-Communist trends in governments or economies has proven so far to be a risky business. With this caveat, trends in environmental policy development and their impact on the activities of the new International Commission will be identified to facilitate forecasts of the future effectiveness of the Convention and programs described in Part Five

10.3.1. MEASURING REFORM WITH INDICATORS

The success of democratic and economic reforms in the CEE countries has been widely monitored. Six of the Danube countries have signed EU association agreements and are judged in a recent report on their approximation to both goals. The "state of their readiness for integration" is evaluated on the basis of the success of their political reform, compatibility of their reformed legal systems with western standards, foreign and security policy orientation, economic reforms and revisions in social policies, and general problems with the multiple requirements of the association process (Brusis and Ochman, in Weidenfeld, 1996, pp 11-26).

The report summarizes experiences with recent elections with the comment that the parties in all of the six countries do not yet have the ability to find political compromises. "In many cases they are not supported by electoral camps, or only by those which are rather limited in terms of social structures or political culture. In consequence, they tend to represent the interests of the small clientele only, thus impeding the formation of political will (Brusis and Ochman, in Weidenfeld, 1996, p. 12)." In addition much power is concentrated in the hands of members of former communist parties, successfully blocking or interfering with the reform process.

The new civil society has problems as its proper role is not widely recognized, particularly with respect to non-governmental organizations. The media remain under government control although new newspapers are appearing in significant numbers. Where foreign companies introduce television channels they do not attempt to use them to control public opinion. Support does continue in all countries for integration into the EU and NATO.

10.3.2. ECONOMIC REFORM

Economic performance has improved but the report stresses the need to evaluate its improvement in comparison with the 1989 GDP of the countries. Slovenia is the first among the six Danube countries to surpass its 1989 GDP level. All countries had stabilized their currencies with the exception of Bulgaria which continued to experience significantly high inflation rates. Unemployment is down but a base of long-term unemployed in rural and weak urban sectors is developing. Structural reforms continue as all countries but Bulgaria and Hungary balanced their budgets. Bulgaria has a critical foreign debt problem.

Consolidation and restructuring of the large state-run industrial enterprises are going slowly, limiting foreign and other investment in them. Hungary remained the most attractive country by far for foreign investors. Between 1989 and 1995 capital in the value of US$10.634 billion, i.e. more than 30% of its 1994 GDP, was invested in the country (Brusis and Ochman, in Weidenfeld, 1996, p. 21). Telecommunications, road and rail networks, continue to need modernization in all countries. All countries are attempting to prevent the impoverishment of low-income groups, to reform health and retirement insurance to make it contribution based, and shifting such functions to private providers. Poverty, unemployment, and crime continued to increase

10.3.3. ENVIRONMENTAL POLICY

With respect to the environmental standards the EU applicants are expected to reach, it is suggested that they do not receive much attention. "...[C]ooperative environmental policy may well be considered 'very important' for integration in dinner speeches (Brusis and Ochman, in Weidenfeld, 1996, p. 25)."

A profile is presented for each of the countries participating in current international cooperative agreements to protect Danube Basin water resources. The brief summaries of the political and economic changes which have taken place are more factual than analytical. Data cover the period from 1990 to 1995-1996. Statistical economic indicators, as opposed to political, or environmental, have been in use world wide for a number of years. Their application to the transition years (Tables 10.2 - 10.3) despite the lack of compatible data for some indicators helps to gauge reform progress in the Central and East European and CIS countries. The political changes countries make to meet the EU criteria for democratic institutions are not easy to measure statistically, but they do yield to analytical formats (Table 10.1).

The summaries in these chapters have received considerable help from transition watchers in the EC and elsewhere as they monitor the progress of applicant countries reported (Weidenfeld, 1995 and 1996; EBRD, 1994-1996).

Indicators which evaluate national and international environmental policies are as difficult to develop as political reform indicators. To be useful in policy making more is needed than the traditional sets of indicators which combine summaries of the ambient concentrations of pollutants and match them with legal standards or recommended criteria. Such data, even presuming accuracy and compatibility, must be related to other variables reflecting population growth and economic development to aid public policy analysis (Hammond, 1995 and World Bank, 1995). Data measured in this way are not yet available.

10.4.1. THE EVOLUTION OF NATIONAL TRANSITION POLICIES

End of Soviet Domination

The country profiles which follow track the general political and economic developments which occurred during the first six years of the post-Cold War transition,

1990 to 1996. Soviet institutional implants, tolerated over four decades, were increasingly rejected by body politics in the 1980s, as east European economies faltered and wavering political linkages were shaken by Gorbachev's reforms. By November and December of 1989 the Communist governments of Bulgaria, Czechoslovakia,

Table 10.1. Economic, Social Indicators for the Danube River Basin Countries, 1990-1995, Austria

INDICATOR		1990	1991	1992	1993	1994	1995
Social Indicators							
Population million		7.7	7.8	7.9	8.0	8.0	8.0
Population growth rate %		0.3	0.3	0.7	0.6	0.5	0.4
Infant mortality rate per 1000 live births		7.8	7.5	7.5	6.5	6.3	5.4
Life expectancy	(M)	73.0	74.0	73.0	73.2	73.4	73.7
	(F)	80.0	81.0	80.0	79.8	80.0	80.3
Literacy rate %		99.0	99.0	99.0	99.0	99.0	99.0
Economic Indicators							
Overall Economic Conditions		1990	1991	1992	1993	1994	1995
GDP in $ bn		158.4	165	186.2	182.1	171.2	197.5
Real GDP growth %		4.2	2.9	1.8	-0.1	3	2
GDP per capita in $		14,500	20,985	18,000	17,000	17,500	20,998
Unemployment rate		5.4	5.8	5.9	6.8	6.5	6.8
Consumer price inflation %		3.3	3.3	4.1	3.6	3	2.2
Exchange rate (av) Sch:$		11.37	11.68	10.99	11.63	11.42	10.08
Current account $ bn		1.2	0.1	-0.6	-0.8	-2.5	-4.5
External debt in $ bn		11.8	NA	NA	16.2	21.5	NA

The communist governments of Hungary and Romania had fallen in the aftermath of large public, and occasionally violent, demonstrations. By mid-1991 Slovenia, Croatia, and Bosnia-Herzegovina had declared their independence from the Yugoslavian federation leaving Serbia part of a two-nation federal remnant; Moldova and Ukraine also had become independent republics.

At the end of the period of Soviet domination each country looked to the west for financial help to rebuild severely damaged economic infrastructures. Each also sought, with varying levels of interest, a replacement for the international security system of the Warsaw Pact. And as they struggled to achieve these two goals they also sought help with changes in political institutions.

10.4.2. CONSTITUTIONAL AND INSTITUTIONAL CHANGES

Early in the transition most countries adopted new constitutions emphasizing protection for human rights, a democratic form of government with multi-party systems and

Table 10.2. Economic, Social Indicators for the Danube River Basin Countries, 1990-1995, Germany

INDICATOR		1990	1991	1992	1993	1994	1995
Social Indicators							
Population million		78.8	80.0	80.6	81.2	81.4	81.6
Population growth rate %		0.5	0.4	0.5	0.4	0.4	0.6
Infant mortality rate per 1000 live births		6.0	7.0	7.0	7.0	6.5	6.0
Life expectancy	(M)	73.0	73.0	73.0	73.0	73.2	73.3
	(F)	81.0	79.0	79.0	79.0	79.6	79.8
Literacy rate %		NA	99	99	99	99	99
Economic Indicators							
Overall Economic Conditions		1990	1991	1992	1993	1994	1995
GDP in $ bn		1,157.2	1,686.7	1,723.8	1,703.8	1,752.6	2,122.1
Real GDP growth %		5.5	4.3	2.2	-1.2	2.9	2
GDP per capita in $		14,600	16,700	17,400	16,500	16,580	NA
Unemployment rate		7.2	6.3	6.7	8.2	9.2	11
Consumer price inflation %		2.7	3.5	4	4.2	2.7	1.8
Exchange rate (av) DM:$		1.62	1.66	1.56	1.65	1.62	1.43
Current account $ bn		46.2	-18.7	-21.7	-15.6	-21.7	-17
External debt in $ bn		NA	NA	NA	NA	NA	NA

market economies. They provided major changes in political processes. The degree to which constitutional changes were supported by legislation, and its implementation, are reflected in the criteria developed by the European Union (EC, 1995 and Weidenfeld, p. 10). These standards were conveyed to the applicants for EU membership at a strategy session held in Essen by the European Council in December 1994.

EC White Paper

The EC's White Paper of March 5, 1995 set forth the dialogue which would take place between the associated countries and the Commission. Institutional reforms were to include a system of legal and commercial rules embodied in extensive new legislation, development of economic institutions such as firms, markets and regulatory bodies, and the creation of a appropriately qualified workforce to achieve these changes. Progress toward privatization was a crucial aspect of the reform. The EU applicants were warned that legislation alone would not be the answer. They must also assume the "important and much greater task of developing adequate administrative and judicial structures, as well as institutions in the business and private sectors" to ensure their effectiveness (Herrnfeld, 1996, p. 14).

The Introduction of the "Civil Society"

The Danube countries which had been part of the eastern bloc or republics within the USSR were influenced by calls for change from organizations within their countries as well as pressures from the west. The "civil society" makes a fundamental change in decision processes since it institutionalizes, democratizes, and protects public participation in the political decision process. The following profiles track the three major sectors: political parties, their structures and goals; the media (press, television, radio), their degree of freedom and the extent of their audiences, and non-governmental organizations representing a spectrum of interests in the public, nonprofit and for-profit sectors. Under reform single parties were to be replaced with multiple, greater freedom given to the media, and non-governmental organizations which had been identified with interests but did not truly represent them replaced by groups more responsive to constituencies.

10.4.3. ECONOMIC POLICIES

Pre-transition economic policies varied from country to country. Hungary, for example, had begun in 1968 to reform its economy along western market lines and to reinforce the change legislatively. Other countries experimented from time to time with liberal measures, particularly during the recession years of the 1980s. Reform is evaluated by privatization levels, market and trade liberalization, revisions in financial institutions including the banking and securities markets, and the role the state plays in taxation and social security. Changes are measured institutionally and by direct application to agriculture, energy, communications, and social conditions. Four major economic criteria for integration are income per capita, the inflation rate, balanced budget, and the extent of trade with Europe.

A number of specific achievements have been noted. All associated countries in central and eastern Europe have a basic form of company law which includes equal treatment of foreigners and foreign companies. Financial reporting following EU requirements is also in place. Priority has been given to the adoption of basic legislation in the financial services sector. Progress has been made in indirect taxation and customs legislation; all of the countries have introduced the value added tax. Industrial standards are also being harmonized with western norms. Agriculture, transportation, energy, and telecommunications are major subjects of reform.

There have been many problems with the legislation required for such substantial changes. Most of the recent legislation has yet to be translated into EU languages. "In all cases," however, the new associate countries "have begun to make a systematic effort to draw up programs for legislative approximation, to set priorities and to verify conformity with relevant Community law (EC, 1995)." Assistance has come through the Phare programme and includes consultation with and technical help to each country.

10.4.4. ENVIRONMENTAL ISSUES AND POLICIES

Austria and Germany

The reach of the Danube that flows from Donauschingen in Baden Württemburg to the Austrian-Slovak border, while heavily dammed, has a relatively low level of pollution. The reduced impact of the waste discharges of Austria and Germany on the river from industry, agriculture, and municipalities reflects stringent emission value requirements and other environmental regulations. Each country has supported the construction or modification of waste water treatment plants and other water quality projects requiring a high level of investment. And each country has had strong "green" movements which supported such regulations and investments but remain critical of the water resource projects, primarily dams for hydro-electric power, which proliferate on the main river and tributaries. Both countries are members of the European Union.

The Danube Program's Strategic Action Plan list of "hot spots" identifies none for Germany and Austria, "due to their longstanding water quality legislation and investment in municipal and industrial waste water treatment facilities (EPDRB, 1995, p. 36)." Germany has occasional increased levels of contaminants and nutrient loads which are being adequately treated. Austria has problems with ground water contamination in its eastern and southern regions because of intensive crop production and cattle farms. Measures are under way to treat these problems (EPDRB, 1995, p. 36).

There is no need to reiterate here the political and economic development of Austria or Germany. (Tables 10.2 and 10.3) In 1990, following the collapse of its communist government, East Germany voted for unification with the Federal Republic. Reunification has necessitated adjustments in economic, political and environmental policies. Austria and Germany are both signers of the Danube Convention and have participated fully in the Danube Program. In particular, an Austrian government official, Wilhelm Kittinger, has effectively led all participating countries in negotiations for agreement on the Convention.

Evaluating Environments

The Danube countries have reported on their respective environmental polices and status in yearbooks, national reviews, and in other ways primarily under the guidance of Danube Program activities. The Program has also supported the harmonization of data monitoring and sharing in the basin so that comparisons are possible and warnings about accidental spills effective (Lack, 1996). As a first step toward that goal, all participating countries provided national reviews of their environmental status, following a common, specific outline. Reports provided information on environmental conditions and policies, particularly with respect to water resources.

The environmental profiles are based on these data supplemented by a number of other reports. The excellent synthesis of the national reviews and complementary data by a team headed by Edward Maltby and supported by the IUCN, the World Conservation Union, has been particularly helpful to this book (IUCN, 1994). Other reports came directly from the countries themselves.

Lack of Compatible, Trend Data

In most of the Danube countries in transition comparative water quality trend data are not available. Harmonization of such important elements as the choice of parameter, methods of sampling and laboratory analyses, development and enforcement of standards and emission limit values is still under way. Water quality data varies throughout the basin, between and within sub-catchments and among countries. Differences in classification schemes and other criteria continue to make basin-wide comparisons of water quality difficult.

Sources and Effect of Pollution

The profiles summarize the unique development of environmental policies in each country. They begin with a statement of major problems inherited from the socialist period. The planned economy under Soviet domination emphasized under-pricing of energy and raw materials which resulted in the waste of these resources. Heavy industry used low-quality fuels, primarily lignite, which resulted in increased levels of air pollution. Intensive agricultural development aided by pesticides and other chemical agents raised contaminant levels in soils, water resources, and food. Price distortions and subsidies discouraged investment in efficient, environmentally protective plant maintenance. Technologies stagnated and produced high levels of waste. Much residue from such waste was left untreated in land fills which polluted the atmosphere and seeped into ground water.

Increased availability of data during the transition period from a number of sources continues to reflect unfavorable impacts on natural resources. While water quality is emphasized in this study, the impacts on natural habitats through decreasing biodiversity, and air and soil contamination are also highlighted. Pending the operation of the international network of monitoring sites (Chapter 22) sources of data are limited. Sources of pollution are now subjected to closer monitoring and wider sharing of data, providing the basis for improving treatment of industrial, municipal and agriculture discharges into surface water and ground water, leaching from solid waste disposal sites and from the effect of engineering structures and their operation.

Industrial enterprises and the resulting areas of pollution clustered along Danube tributaries, resulting in "hot spots," local land areas, stretches of surface water or specific aquifers which are subject to excessive pollution from an identifiable source and which require particular action to prevent or reduce the degradation caused (EPDRB, 1995, p. 36). Point source contaminants have created pockets of air, water, soil, and biota pollution, particularly in or near urban areas. A survey of health concerns and the areas which have been identified as top priority for clean-up conclude is followed by a discussion of the environmental policies which have evolved in the early transition years.

Health Questions

Available data on the effect of pollution on health has been scattered, with much of it relying on country reports from the pre-transition period. A recent study makes an important contribution to a better understanding of basin health issues by providing a rationale for the collection of data (Chorus, Ingrid, 1995). It points out that the Environmental Health Action Plan for Europe (EHAPE), adopted in 1994 (Helsinki), reported that the overall incidence of waterborne infections is difficult to quantify because medical advice is sought only in the most serious cases, such as cholera, typhoid or hepatitis. Gaps in information require well planned and systematic research, especially in epidemiological problem areas.

Research reveals the types of diseases key to Danube River Basin water pollution. Infectious diseases resulting from microbial contamination is evident throughout the river system, with heavy pollution by fecal bacteria and viruses in most of its reaches (Chorus, 1995, ii). While the exact incidence with respect to severity and time is not known, numerous cases of dysentery, jaundice, hepatitis and infections of the intestinal tract such as salmonella, streptococcus, Giardia Lamblia, have been documented. Cholera has also been reported. Infant methemoglobinemia, more widely known as blue baby syndrome, is second in importance after infectious diseases. New-born infants may be deprived of oxygen because of the effect of nitrate concentrations in drinking water. Extensive use of nitrates as a fertilizer is responsible for such concentrations. Remedial measures include the distribution of bottled water for infants and waste treatment. Other chemicals used in agriculture and industry are known to contribute to disease.

Drinking water contamination is the major source of these diseases. The cure is to increase the available treatment of water supply sources. Areas with a well developed and maintained central water supply system have far fewer cases of water-borne infections. Water shortages are increasing in many Danube areas, hampering personal and household hygiene, and encouraging the storage of water by individual households which frequently results in bacterial contamination. The food supply may be contaminated by pathogens transmitted by irrigation water. The data for levels of industrial contaminants from industrial sewage or leaching from landfills are "fragmentary" but the hot spots identified in the Strategic Action Plan (see below in each Country profile) speak to the lack of treatment for a variety of such sources.

Government Policies

The strengthening of democratic institutions is an item on the Danube Program agenda, as international funders agreed that concepts and processes would have to be changed if environmental policies were to approximate those of western Europe.

A study sponsored by the Regional Environmental Center (REC) in Budapest points out the differences between east and west European legal systems as they "reflect both the concept of law and mechanisms created to use and enforce the legal acts (Bandi, 1996)." One essential difference was the omission of the role of the public in passage and enforcement of legislation in the east and its inclusion in the west. In addition, laws setting environmental quality standards were usually not supplemented with effective implementation mechanisms. They did not include, for example, economic incentives to ensure compliance by subsidized state owned businesses. The regulators and regulated were agencies of the same ruling party. Thus concepts about the role of law had to be changed; legislation is a question of content and strict enforcement.

One of the most difficult tasks in writing effective regulations is to establish conditions and mechanisms which ensure clear rules for administrators to follow and equal opportunities for all participants to be heard. The REC study reviews compliance legislative content presented are more helpful.

The REC study offers insights into impediments to the easy adoption of western environmental policies by the new democracies of eastern Europe. Other more urgent issues take precedence. More and better guidance is needed to help legislators understand the EU legislative process and more skilled personnel is needed in each country to help with drafting legislation. Cost-benefit studies would supplement the legislative process with accurate estimates of the cost of adopting and not adopting environmental protection policies.

Table 10.3 Country Share of Danube River Basin Area

Country	Area km^2	% Country	% Basin
Austria	79.5	96	10
Bulgaria	48.2	43	6
Croatia	37.0	65	5
Czech Republic	25.0	30	3
Germany	59.6	17	7
Hungary	93.0	100	12
Moldova - Prut	12.0	4	2
Romania	232.2	98	29
Slovak Republic	46.5	97	6
Slovenia	25.8	53	3
Ukraine	29.5	4	4
Bosnia, Serbia	112.7	NA	14
Switzerland, Italy, Poland, Albania	5.3	NA	1
Total km^2	806.3		100

10.5. COUNTRIES PROFILED

The profiles which follow concentrate on the changes which have taken place in the political, economic, and environmental policies in the former communist states in the first transition years. All but two of the nine countries in the Danube catchment area which were once counted among the eastern bloc, the former Yugoslavian countries of Bosnia and Serbia, have participated in the Danube Environment Program. Six have applied for EU membership and are NATO partnership countries. The inclusion of Serbia and Bosnia in Danube cooperative activities is anticipated. By 1997 Bosnia was taking part in the Danube Program, leaving only Serbia a non-participant. The two former Soviet republics in the basin, Moldova and Ukraine, are NATO partnership countries and have been initiating changes in political, economic and environmental policies to approximate western norms.

CHAPTER 11

Bulgaria

11.1. Fall of Zhivkov

Todor Zhivkov, Bulgaria's Communist party and government leader for 35 years, was ousted in November 1989 and replaced by the more moderate foreign minister, Petar Mladenov. Demonstrations took place in Sofia demanding elections for a new government, respect for human rights and the environment, and an end to police brutality. After a large rally on December 10, Mladenov proposed that the Communist party's monopoly on power be ended and that elections be held in June. Andrei Lukanov, the last communist party prime minister, remained in power until November 1990. Leaders of the former Communist party renamed it the Bulgarian Socialist Party (BSP) and won a small majority in the June 1990 election for the Grand National Assembly which was to prepare a new constitution.

11.2. Political Reform

11.2.1. ELECTIONS

Parties

In August 1990, Zhelyu Zhelev, a member of the reform coalition, was elected to the presidency, an office with very little power, by the parliament. In November 1991 Filip Dimitrov formed a government of parties represented by the Union of Democratic Forces (UDF) following its strong showing in the second parliamentary election held in October. Zhelev was confirmed as president in a popular vote held in January 1992. Dimitrov lost out in a no-confidence vote in October 1992. The ruling coalition continued to have problems with its coalition parties as they tried to maintain agreement on important issues. The 1994 elections resulted in support for Zhan Videnov from the BSP as prime minister. Leaders of the UDF claimed that the Communists had returned to power. By the fall of 1996 it was evident that the BSP government under Videnov had failed to meet the challenge of economic and financial reform. The country was experiencing severe hardships; polls showed that 60 percent of the people regarded the BSP government as a failure (Deimel, 1996, quoting a BBSS Gallup poll in Trude 6.7.1996, p.12, p. 29).

An Assassination
In November 1996 Petar Stoyanov, the vice-chairman of the UDF, was elected to the presidency, winning 59% of the vote in a run-off election with Ivan Mazarov, the candidate for the governing BSP. A factor in the election was the strong showing by George Ganchec of the Bulgarian Business Bloc who polled 23% of the votes in the first balloting. On a somber note, Lukanov, the former prime minister, was murdered, a victim of a gangland-style shooting in front of his apartment building in Sofia, shortly before the elections, on October 5, 1996. Lukanov had been invited to the Woodrow Wilson Center in Washington as a visiting scholar to begin in October. He is said to have been critical of the leadership of Prime Minister Zhan Videnov and the economic policies of the BSP (Perlez, *The New York Times*, November 3, 1996).

Instability Prevails
A two-day special congress of the Bulgarian Socialist Party was called following the loss of the presidential election by its candidate, Mazarov. On December 21, 1996, Prime Minister Videnov, announced that he was resigning from office along with his cabinet. The party chose a new leader who would presumably take the place of Videnov until the next election. The future was bleak for the party leadership and the country as Videnov and some of his cabinet functioned as caretakers pending a new election. The IMF was still demanding reforms before it provided the next payment of a loan of about $600 million. A strategy paper drawn up in May 1996 outlined program goals for reform by 1998 with the banking sector as a special concern. The BSP had failed to privatize large state enterprises, reform the banking system or remove currency control from the national bank.

Large demonstrations continued to be held, primarily in Sofia, as opposition leaders demanded new elections following the resignation of Videnov. Crowds supported the holding of immediate elections in advance of the 1998 date. A compromise appeared to have been reached early in 1997 but was later abandoned.

Uncertainty in International Cooperation
Bulgaria continued its participation in the various environmental activities of the Danube Program but clean-up of polluted sites had not been a popular issue since the early days of the overthrow of the Communist government. Initial support for greater protection of the environment diminished as the country came to grip with severe economic problems. Although the Ministry of Environment gave support to the Danube Program, environmental issues in the political and economic transition were not emphasized during the early transition years.

11.2.2. WESTERN ORIENTATION

Bulgaria's EU Association Agreement went into effect in February 1995. Since March 1995 government officials and EU representatives have been working together to prepare Bulgaria for entrance into the Union. Bulgaria has received assistance from IMF and the World Bank since the early nineties and has been a member of the UN for over 40 years. The country has also entered into a NATO Partnership for Peace agreement. Bulgaria's major foreign policy goals are western oriented. Its leaders from time to time have wanted to play a lead role in the formation of a regional cooperative movement which might lead to a Balkan Helsinki accord. The country is party to a number of international conventions and participates in a Black Sea regional economic cooperative activity initiated by Turkey in June 1992. Foreign policy has generally favored a "non-alignment" strategy based on bilateral treaties and the normalization of relations with both Greece and Turkey.

11.2.3. CONSTITUTIONAL AND INSTITUTIONAL CHANGES

The constitution adopted in July 1991 by the National Assembly established a parliamentary republic with the usual guarantees that it be governed by rule of law, preserve political pluralism and a free enterprise system based on private property. Basic human rights were protected as well as a safe environment. Regional and local government reform has included the reorganization of regional administrations. In the reorganization, 35,000 government employees were said to have lost their jobs.

The Civil Society Component

The post-Cold War political party system in Bulgaria has been described as a "very delicate constellation of powers where no political power can hold strategically important positions for a long period of time (Dimitrov, 1995, p. 23)." Stalemate over the major political and economic decisions has been the result. "Cross party" governments have been ascendant as parliamentary units group and regroup. Koulov identifies a "bi-polar" party structure despite the presence of over 100 smaller parties, with the BSP and the Union of Democratic Forces (UDF) representing opposing political positions. The Turkish Movement for Rights and Freedoms (MRF) has elected members to the three parliaments (Koulov, 1995, p. 2). Immediately after the ouster of Zhivkov there were 17 reform parties in the UDF coalition.

The third and most recent free elections, for parliament, took place in December 1994, when the Socialist party, in a coalition with a left-wing agrarian party and a splinter political wing of the NGO Ecoglasnost, a pro-environment organization, won with an absolute majority (125 of 240 seats). The reform coalition which had won near majorities in the 1991 election took only 69 seats. Three social groups have been identified as the primary supporters of the Socialist Party. Small-scale farmers fear radical land reform and a return to private ownership. Residents in smaller towns react to perceived reductions in their standard of living. Pensioners on fixed incomes believe they have been left out of the new market system. Support for the former Communists is based on self-interest rather than ideology.

Media

Like some other east European countries, a large number of daily and weekly newspapers claim the attention of the reading public. There are now about 500 daily newspapers, most of them privatized. Ten leading newspapers account for about half of the total circulation of daily papers. Two state-run television channels continue to operate and have been joined by one privately owned channel. Broadcast journalists were dismissed from the Bulgarian National Broadcasting Company, the only national radio network, in 1995 for criticism of the government (Deimel, 1996, p. 32).

Interest Group Representation
The non-governmental organizations which had previously served as government claques have gradually evolved as representatives of major groups such as trade unions, protectors of natural resources, education, and so forth. The leading confederation of independent labor syndicates has been reorganized along industrial lines. A smaller confederation has been strongly anticommunist but has left the UDF although it was a founding member of the coalition. New private-sector employers have yet to form effective groups although an organization which calls itself G13 and includes thirteen private entrepreneurs and bankers is attempting to attract support for improved government market-oriented policies.

11.3. Economic Policies

Bulgaria's economy shared the sharp decline experienced elsewhere in eastern Europe in the 1980s. As a result of the failure to invest in and upgrade its industrial sector production of consumer goods declined. It had been more heavily dependent on CMEA trade than any other east European country and had also accumulated a sizable foreign debt. Problems facing the new government in 1990 were severe. Basic products such as sugar, cooking oil and soap were in short supply and interest could not be paid on a foreign debt of $10.6 billion. Under-pricing under Soviet domination caused a higher use of energy and raw materials.

Pollution levels were exacerbated by an emphasis on heavy industry and the use of low-quality fuels, primarily lignite. Technologies stagnated and produced significant levels of waste. Price distortions and subsidies discouraged investment in efficient, environmentally protective plant maintenance.

11.3.1. WESTERN AID

Reforms which would support institutional progress toward a market economy were stimulated by the terms of a 1990 IMF loan. A macro-economic stabilization program effective in February 1991 included price liberalization, establishment of a market determined exchange rate, a check on wage growth, a restructuring of the budget with strong requirements for budget deficit reduction, and a tight credit policy. A revised commercial code was in place in May 1991. Laws on property and the use of agricultural land, as well as the liberalization of trade, the addition of a value added tax, and other legislation were passed. Initial reform legislation has faltered; its implementation characterized by indecision and inconsistency (Wyzan, 1994, 540 ff).

Table 11.1. Economic, Social Indicators for the Danube River Basin, Transition Countries 1990-1995, Bulgaria

INDICATOR	1990	1991	1992	1993	1994	1995
Social Indicators						
Population millions	9.0	8.6	8.5	8.5	8.4	8.3
Population growth rate %	-0.3	-0.2	0.5	-0.4	-0.3	-0.3
Infant mortality rate per 1000 live births	14.8	16.9	15.9	15.5	12.0	11.4
Life expectancy (M)	69.0	69.0	69.0	69.6	70.0	70.4
(F)	76.0	76.0	76.0	76.3	76.7	77.1
Literacy rate %	NA	NA	98	98	98	98
Economic Indicators						
Overall Economic Conditions	1990	1991	1992	1993	1994	1995
GDP in $ bn	50	5.5	8.2	9.2	9.6	11.1
Real GDP growth %	-9.1	-11.7	-7.3	-2.4	1.4	2.5
GDP per capita in $	NA	NA	3,800	3,800	3,830	NA
Industry % of GDP	51	48	43	42.3	NA	NA
Agriculture & forestry % of GDP	18	15	16	13	NA	NA
Unemployment rate (in % of labor force, end-yr)	1.5	11.5	15.6	16.4	12.8	14.2
Consumer price inflation %	21.6	333.5	82	72.8	96.2	62
Exchange rate (av) Lv:$	2.4	18.4	23.3	32.7	66	70.7
General government balance in % of GDP	-12.8	-14.7	-15	-15.7	-7	NA
Trade balance in $ bn	-0.8	0	-0.2	-0.9	0.2	0.4
Current account $ bn	-1.2	-0.9	-0.4	-1.2	0.2	0.1
External debt in $ bn	10	12	NA	12	12	9.4

Financial assistance to Bulgaria has included the initial 1991 IMF loan and a subsequent loan of $700 million. A consortium of international funders cover Bulgaria's foreign debt of $8.2 billion. The World Bank made a structural adjustment loan in August 1991 of $250 million; the second installment was not offered until 1994 because of the Bank's dissatisfaction with Bulgarian failure to abide by its criteria. The country has received support for projects related to agriculture, the environment, energy, telecommunications and the financial and health sectors averaging about 90 million ECUs per year. Humanitarian grants have been received and special loans from the European Investment Bank for small and medium-sized projects for improving energy supply, telecommunications, air traffic, and transport.

11.3.2. MEASURING CHANGE

Macro-economic adjustments to reform policies have been under way since 1990. Some success has been achieved as a result of these policies, for example a reduction of state intervention in prices of goods as well as substantial increases in the price of electricity and other energy sources. Wages were maintained or reduced and enterprises given the legal power to fire employees. In 1992 according to an OECD report "the monetary overhang was eliminated; prices and trade were liberalized; hyperinflation was avoided; budgetary adjustment was undertaken; wages were kept under control; and external balance was obtained (quoted in Wyzan, 1994, p. 540)." These and other optimistic appraisals were confronted almost immediately with the harsh realities of a steep decline in agricultural and industrial production accompanied by high unemployment, continuing budget deficits, uncomfortably high inflation, and excessive volatility in exchange rates (Wyzan, 1994, p. 531).

Projections for the future remain poor. Agriculture and the food industry are not advancing and the only sectors meeting expectations have been services and tourism. Privatization has made little headway. The gross domestic product has declined sharply, accompanied by high inflation and unemployment. While legal reforms provide the same rights and duties to foreigners as Bulgarians except with respect to the ownership of agricultural land, progress in privatization has been limited. Ninety percent of manufacturing enterprises are still owned and run by the states. "As a location for foreign investment Bulgaria is still ranking last behind the other associated countries (Dimitrov, 1995, p. 40 and Deimel, 1996, p. 43)."

11.3.2. INDEBTEDNESS PROBLEMS

Domestic and foreign indebtedness remain a problem. The state-owned companies owe large sums; an estimated 25% of government expenditures are needed to cover such loans. The rescheduling agreement reached with the Club of London in 1994 covers $6 billion in debts and $2 million in interest, reducing the overall debt by half. One billion dollars had to be paid in 1995 alone to creditors. The agreement has helped to make it possible for Bulgaria to gain access to financial help from other quarters.

11.3.3. OTHER ECONOMIC FACTORS

Almost all agricultural land collectivized under socialism has been restored to pre-1947 owners or their families. The resulting division into a large number of economically unprofitable small holdings has sharply reduced food production. The institution of cooperatives is in progress.

Bulgaria has problems with energy supply. As recently as 1994 45% of its electricity came from nuclear reactors at Kosloduj. The World Bank and the European Bank for Reconstruction and Development have urged the phasing out of a reactor block in 1997. This source cannot be replaced with coal. Bulgaria has received 200 million ECUs from EIB to modernize its energy sources. Its telephone and transportation systems need substantial improvement.

From 1990 to 1995 the number employed has dropped from 3.9 to 2.1 million. Unemployment is probably substantially higher than the government figure of 12.8%. Economic reforms have effectively cut the real purchasing power of the population by more than half. The UN sanctions resulting from the Serbian invasion of Croatia and Bosnia have closed down profitable transit traffic for Bulgaria and resulted in estimated losses of from $6 to $8 billion.

11.3.4. OUTLOOK

Economists are generally pessimistic about Bulgaria's economy. Privatization of large-scale industrial enterprises has made less progress than in any of the other former CMEA countries. Wyzan reports:

> The banking system is still dominated by state commercial banks with portfolios of uncollectible debt from struggling state enterprises, and the recent bad-debt deal will not solve this problem. Even the achievements in economic stabilization are fragile. In the medium term the need to service the obligations under the London Club agreement will represent a heavy burden for the economy. Growing interest payments on domestic debt make the fiscal situation worrisome, and exchange rate crises like those of late 1993 and early 1994 are far from impossible. In the end the biggest Bulgarian economic problem is undoubtedly the politicization of all major policy decisions (Wyzan, p. 550-1).

In April 1996 *The Economist* presented a picture that was even bleaker than that described above. Bulgaria's banks "have indulged in a mixture of ill-judged spending, over-expansion and corruption that would make even some Russian bankers blush. Mixed with lax supervision, inadequate regulation and industrial chaos (most large companies do not bother to service their debts) this cocktail has put the country's banking sector flat on its back (*The Economist*, April 20th, 1996, p. 66)." By October, a January deadline for a payment of $130 million to the London club of commercial bank creditors, and for payments of a $350 million bill for imported fuel and grain loomed (Perlez, *New York Times*, October 28, 1996, A6). In September Bulgaria's central bank had propped up the lev by tripling its base interest rate to 300%; it also took over nine banks in the face of a collapsing economy (*International Herald Tribune*, September 25, 1996, p. 15).

11.4. Environmental Issues and Policies

While Bulgaria, like the other former socialist countries, lacked protective environmental policies it had less overall industrial and mining development and therefore did not experience environmental destruction on a level with other east European countries, in particular the Czech Republic. Rather, point source contaminants created pockets of air, water, soil, and biota pollution, particularly in urban areas. Studies before and since the change in government indicate that some of this contamination poses serious threats to health in many parts of the country. The discussion of the current environmental situation in Bulgaria which follows depends on various reports by international agencies, in particular the one prepared with the help of the World Conservation Union (IUCN, 1995)

11.4.1. MAJOR DANUBE ENVIRONMENTAL ISSUES, EARLY TRANSITION

Bulgaria is a predominantly mountainous country, which divides into three parallel and approximately equal east-west zones: the Danube plain in the north; the Balkan Mountains in the center, and the Thracian Plain and Rhodope and Pirin Mountains in the south. Its major watersheds are three in number: an area drained by the Danube through tributaries; a coastal area which drains through moderate-sized rivers and inland seas into the Black Sea, and a southern section which drains into the Aegean Sea via Greece and Turkey.

Multiple Problems

Bulgarian environmental problems are evident in the pollution of its water, air, and soil and the growing aridity of its northeastern section. Information and data about such conditions are limited and comparisons with other countries remain difficult until the Danube Program monitoring program is in place.

While there is substantial recovery of water from industry and municipal use, most recovered water is too polluted for use. Long term strategy for water supply relies on an increasing amount of water intake from the Danube. One third of water supply comes from surface water and the remainder from ground water. Excessive use of ground water, however, has created availability problems. Several years of drought conditions in the northeastern part of the country have caused serious depletion of groundwater sources. More hydrogeological studies are needed to locate aquifers and their infiltration areas.

Water Quality Monitoring

A water quality monitoring system has been in use in Bulgaria since 1975. Waters are classified into four categories and standards adopted for each from (1) pure to (4) strongly polluted. Treatment has lagged on all major rivers including the Danube tributaries. An estimated fifty per cent of the ten major rivers in the country are seriously polluted. Industrial plants, city sewage and feedlots which discharge untreated streams of chemicals, heavy metals and sewage effluent into rivers are the main sources of pollution.

Monitoring of eight or nine heavy metals plus bacterial levels has taken place over a period of several years although its accuracy is in question. A widespread problem in drinking water in some areas has been the level of nitrates. More than 30% of the districts show levels above the standard, according to a World Bank study (Hertzman et al, 1995, p. 41). Nitrates in drinking water from inadequately maintained or designed or improperly located rural septic tanks, feed lots and other agricultural enterprises are especially harmful to infants. The lack of evidence that measures were being taken to control methemoglobinemia (the "blue baby" syndrome which results from lack of oxygen), including supplying bottled water for infants, was "quite startling" to the World Bank study team.

Other Resources

There is a general trend toward a decrease in biodiversity along the banks of the Danube which could be attributed to the effects of the Iron Gate dams and a general increase in nutrients and organic matter from point and non-point sources. Biodiversity is especially diminished in downstream tributaries as the result of industrial and urban pollution. Wetlands distribution has been greatly reduced from the turn of the century, primarily as the result of agricultural drainage. Some wetlands are under special protection; Bulgaria has six national parks within the Danube drainage basin. The Strategic Action Plan lists five sites which need to have their natural purification and assimilative capacity reestablished (EPDRB, 1995).

Bulgaria has experienced significant air pollution in its Danube catchment area. In urban areas sulfur dioxide and dust concentrations have frequently exceeded admissible limits. Degraded and polluted lands are estimated to cover extensive areas in the Bulgarian part of the Danube catchment, having experienced erosion, waterlogging or salinization.

Pollution Sources

Heavy industry, which continues to use outmoded technology, accounts for most environmental pollution from industrial sources. It has been dependent on subsidies for energy and raw material and not yet modernized to achieve a cleaner environment or to make products more competitive in European markets. Compounding serious air and water pollution problems are current industrial waste disposal practices. Wastewater from metals extraction, for example, has been "lagooned" for many years and has been a serious threat to ground water.

Most of the main polluting enterprises in the Bulgarian part of the drainage basin discharge their effluents directly in to the Danube, including the Kozlodui nuclear plant. The latter is one of the largest in eastern Europe. All units are of Russian design; four of them lack safety features presently required in most of the world. The major polluting industrial sectors are pulp and paper production, industrial chemicals, the extractive industries and metals processing, with textiles, leather and construction products also important.

Twelve wastewater treatment plants are operative in the Bulgarian section of the Danube catchment areas but use outdated equipment and experience overloading on a regular basis. Sludge residue contains heavy metals and thus cannot be used for fertilizer.

Agricultural lands constitute 6.8 million hectares of its total land area of 11 million. An estimated 20% of agricultural and forest land is degraded by pollution. Natural causes such as erosion and acidification account for nearly 80% of this but emissions from industrial plants have contaminated some areas with heavy metals concentrations in excess of acceptable levels of concentration. Due to prevailing dry conditions, agriculturally developed land in Bulgaria requires irrigation from surface or groundwater. Irrigation is the largest use of water in the Bulgarian part of the Danube basin, accounting for 70% of total use.

Mineralization through salinity and electrical conductivity is the chief contaminant of irrigation water. Crops are sensitive to nitrate concentrations from animal feedlots and from chloride at levels above standards. Selenium and arsenic have been found to be above recommended maximum concentrations for irrigation waters. There has also been significant impact on agriculture from coal mining waste.

No specific disposal areas for industrial wastes are available. They are mixed with solid household wastes and frequently basic sanitary requirements ignored. Often solid wastes are deposited in the immediate vicinity of river channels. The Ministry of Environment has been preparing laws and regulations for solid waste disposal to address these problems. Nevertheless, the limited financial resources of local communities and industries mean that they are dependent on financial support from international sources. The solid radioactive wastes of the Kuzludui nuclear power plant are sent to Russia for further treatment while contaminated waste water is kept within the plant in special containers.

Engineering Structures

The Iron Gates dams are reported to influence the hydraulic regime of the Bulgarian reach of the Danube as well as its environmental conditions. Problems include daily water level fluctuations of up to one meter during the autumn and winter as well as increased incidence of low flows and riverbank erosion. There have been accidental heavy discharges endangering an area of 200,000 inhabitants. International agreements regulating minimum discharges have not been adhered to so that much lower discharges than required have been frequent. Two new dams are planned to regulate the rate of flow downstream from the Iron Gate dams.

Hot Spots And Health

The Danube Program's Strategic Action Plan (EPDRB, 1995, pp. 37-40) identifies a total of 29 sources in the Danube catchment area which need new or modernized facilities for treating waste or urgent attention of other kinds. Fourteen are considered to be high priority sites. The reservoirs and power facilities of the Iron Gate dams need changes to prevent further damage to Bulgarian river banks. One hundred and thirty kilometers along the Bulgarian bank and 55 km along Danube islands have been severely damaged through erosion and abrasion from dam operation.. There is as yet no estimated cost for this project.

Seventeen of the recommended areas need waste water treatment plants (WWTP); of these eight are high priority. One of the largest is on a major Danube tributary, the Yantra, at Gorna Oriahovitza, a city of 40,700. Ninety per cent of the population is served by a sewer system but the plant itself is ineffective. A new facility, which is expected to be built in 1997, would cost a total of US$23.5 million. Other waste water treatment facilities are planned for sites on the Yantra (Sevlievo), the Danube (Ruse and Silistra), the Iskar (Sofia), and the Ossam (Troyan). A number of manufacturing and processing enterprises which are currently sources of pollution are identified for modernization.

11.4.2. GOVERNMENT POLICIES

Bulgaria's outdated environmental legislation, passed in the 1960s, provided for the establishment and enforcement of environmental standards, among other requirements for the protection of its natural resources. The "Law on the Prevention of Air, Water and Soil Pollution" and the "Law on the Protection of Natural Environment" have been amended several times. They set standards and emission limits (high standards, low enforcement) and authorized other measures to protect land and water resources.

There are many other laws related to the environment; by one estimate, as many as 120. The most significant are laws to regulate mines and quarries, protect forest and plants, regulate the water supply and protect public health, protect cultivated and pasture land, regulate regional and settlement development, game reserves, use of atomic energy and the Bulgarian Maritime Territory. By general agreement among national and international specialists, standards may have been high but enforcement was weak in every sector.

The Bulgarian constitution adopted in 1991 includes three articles supportive of resource protection policies. The duties of the state include a guarantee of environmental protection and rehabilitation; maintenance of biodiversity, and the reasonable use of natural resources. It establishes the right of citizens to a healthy environment and their responsibility to maintain it. The state has the exclusive ownership of waters, forest and parks of national significance as well as natural reserves such as mineral deposits, and littoral beach strips, to which it may grant concessions (Regional Environmental Center, 1996, p. 17).

The environmental strategy adopted under the framework Environmental Protection Act of 1991 set a number of goals for the year 2000. The Act was based on the concept of sustainable development. It set pollution prevention objectives and called for an entire renovation of the regulatory system, particularly for water, and is also supposed to meet the requirements of the new economic conditions and policy. The Act "marks the beginning of a new normative system for environmental legislation, by setting forth objectives, strategies and general normative requirements for the state environmental policy, and also of the powers and duties of state authorities, companies and the general public, in correspondence with their constitutional powers and duties (Regional Environmental Center, 1996, p. 15)." It established the leading principles embraced by EU Council Directives: emphasis on polluter responsibility for the preventing or remedying the damaging effects of pollution principles and the public's right to public information about the status of the environment. To help to achieve ready access to information about the status of the environment a National Center for the Environment and Sustainable Development has been established.

Bills have been prepared and sent to the parliament by the government, including one to implement specific water policies. The government's program includes a commitment to accelerate the approximation of EU environmental legislation, and has identified working groups to proceed with this step. Despite the good intentions of the legislation and those administering it in the government "[t]he remaining environmental laws of the Republic of Bulgaria were adopted long ago. They are outdated and do not meet contemporary social-economic and political conditions in a country undergoing transition towards a market economy and a democratic society (Regional Environmental Center, 1996, p. 16)." No legal framework has been developed, for example, for waste management. The management and disposal of municipal waste depends on local municipalities; standards are neither clearly established nor followed and legislation is needed. The legislation necessary to approximate EU directives is moving through the political system at a slow pace.

11.4.3. NATIONAL ENVIRONMENTAL ACTION PLAN

The Ministry of Environment adopted a National Environmental Action Plan in 1992 based on a two-year strategy study completed with the help of the World Bank and the US AID (Mindjov, May 1995). The plan was updated in 1994 and 1995 and reviewed at a public hearing. Objectives are broad: to strengthen institutions, including legislation, develop a system of financial and economic mechanisms for environmental management and monitoring, and to integrate its environmental policy with the environmental legislation of the European Union. It also provides for sharing information about the environment with non-governmental organizations.

Four priority areas have been identified for the reduction of air pollution. A national priority would seek to minimize heavy metal contamination of drinking water supplies. There are ten key recommendations for policy and institutional development. Bulgaria had not as yet, however, developed or implemented a methodology for identifying priorities for investments in environmental protection. Problems with funding and enforcement are severe and expected to remain so.

Key Institutions

The reorganized Ministry of Environment continues to set and enforce environmental standards. It issues permits and levies sanctions, prepares legislation and data base information, monitors ambient water and air quality; oversees discharge permits and industrial monitoring. Its inspectorates cover various geographical subdivisions from river basins to parts of basins and, in some cases, urban and rural areas. The inspectorates do routine monitoring and sample point sources at municipal and industrial sites and at other designated water resource sites. They maintain laboratories that conduct routine tests on certain designated parameters and issue discharge permits to municipalities and industries. Allowable levels of contaminants are set and these are exceeded, fines are collected.

Since February of 1992 the Ministry has reorganized its central research function as a Laboratory and Information Center with three units: a monitoring section, a laboratory, and an information management section. Monitoring is limited to 20 pollutants. Samples are analyzed in the Center and the results recorded. There are overlapping and confusing responsibilities within organizations with environmental responsibilities. In addition different methodologies are used, yielding incompatible and inconsistent results while equipment and resources are inadequate.

Other Administrative Agencies

There are a number of other administrative agencies with environmental management functions. Water resource use is controlled by the National Water Council. The Ministry of Regional Development has responsibilities for municipal water supplies, sewage systems and waste water treatment facilities. State-managed companies operate water supply and treatment facilities. The Ministry of Health sets and monitors ambient water supply drinking water standards and has special functions for the water quality of household water supplies. The Director of Health Prophylaxis and State Sanitary Control under the Ministry of Health has units for environmental needs, occupational health, epidemiology, nutrition and the health of school children.

Forests and most of the protected areas are managed by the Committee of Forests; the extraction of mineral resources is controlled by the Committee of Geology and Mineral Resources and nuclear safety is under the guidance of the Committee on the Use of Atomic Energy for Peaceful Purposes. The Ministry of Territory and Settlements, Housing and Construction coordinates waste water and supply construction for which there is limited state funding. It oversees the quality of construction, and sets standards for it. Direct responsibility for management is now assigned to local governments.

11.4.4. POLICY OVERVIEW

Enforcement is not the only activity awaiting the adoption of new legislation. Fragmentation of its subject matter requires planning help from a central source which can provide accurate statistics on pollution levels to support projections for future policy decisions. Current planning instruments are not respected and can have little use in the improved environmental management of the nation's resources. Government support for the construction of waste treatment facilities as well as for the adoption of new technology in industrial plants and municipalities is at a low level in the face of its desperate debt situation.

CHAPTER 12

Croatia

12.1. Impact of Civil War

The civil war in the former Yugoslavian Federation, which began in July 1991 when the Serb-dominated Yugoslavian Army attacked Slovenia and Croatia, has had a devastating effect on Croatia. Despite the very large number of Serbo-Croatian marriages and other indications of peaceful coexistence, memories of the past lingered in the former Yugoslavian Federation. Elections following the break-up of the federal Communist Party in January 1990 were won by ethnic nationalists in all countries except for Macedonia. In May 1990 Franjo Tudjman of the Croatian Democratic Union (HDZ), was elected president of Croatia. When his government rejected the demand of Serb minorities for cultural and territorial autonomy, hostilities began. Violent incidents between Croats and Serbs were occurring regularly by the beginning of 1991.

The Yugoslav People's Army, which had its roots in the partisan bands of World War II, was the fifth largest in Europe. Ostensibly the protector of all of Yugoslavia it was dominated by officers who sought hegemony for Serbia. Following declarations of independence by Slovenia and Croatia in June 1991, the Army fought to protect Serb interests and populations in both countries. Full-scale fighting broke out in July. After defeat in Slovenia Serb forces invaded and occupied Croatian territory. Brutal artillery shelling devastated the historic cities of Vukovar, a Danube port, and Dubrovnic, on the Adriatic coast. "[T]he first major war crimes in Yugoslavia since World War II" had been committed (Zimmermann, 1995, p. 7). By early 1992 Serbs supporting secession from Croatia had taken over 30% of its territory. In January Croatia and the Serb secessionist government of Krajina agreed to a truce protected by a UN peacekeeping force. The center of conflict moved to Bosnia. Armed conflict there and in Croatia has left an estimated 300,000 people dead and 3 million homeless.

12.2. The Country under Tudjman

The progress of democratic reforms in Croatia has been slow. A new constitution was adopted in 1989 and a large-scale government reorganization took place in 1994. Reform has been marred by several moves on the part of its president to limit free discussion in the media and to protect one-party rule by the Croatian Democratic Party Union. On November 6, 1996 Croatia was admitted to the Council of Europe, the organization founded in 1949 to promote European unity which now includes 40 members and has been particularly concerned with human rights since 1950. Croatia's admission was delayed for six months over human rights concerns. It has not yet applied to the European Union for associate membership.

12.2.1. LOCAL GOVERNMENT

Tudjman has been criticized for interfering in the election of the Zagreb city council which appoints the city's Mayor. A newly elected council, governed by a majority coalition of opposition parties, took office in October 1995. The council's mayoralty appointee was rejected and in March 1996 Tudjman named the former Minister of Construction Mayor. When she received a vote of no-confidence from the council; he reappointed her. The council itself was dissolved by presidential fiat in May 1996. There have been other interferences with the elected council (Hedges, May 2, 1996).

Critics perceived these actions as a demonstration of the President's growing intolerance and refusal to sanction democratic change. There is a struggle over the sale of state-owned businesses, many of which have ended up in the hands of the President's close associates. According to a diplomat interviewed by the New York Times correspondent, "The ruling party has made a lot of money from the privatization of state properly. Tudjman is not about to turn over the city administration, that can monitor these sales, to his opponents (Hedges, May 2, 1996)." A move to bar coalition rule in the city council was said to be under way. Elections for municipal officials and the upper house of parliament were to be held in March 1997.

12.2.2. THE MEDIA

The one-party government controls all television channels, all but one radio station and three of four national newspapers. A law passed in 1995 made it illegal to criticize the president or other top government officials. Objections to the move by the one-party government to make Croatia a military power was voiced by a retired Croatian general, Antjun Tus in October, 1996. Despite the fact that the station, "radio 101," had helped Tudjman when he was a dissident challenging the Communist authorities, the station has been threatened with the loss of its license (Hedges, October 9, 1996).

When Tudjman's government shut down Radio 101 in November 100,000 people demonstrated in Zagreb's old town area. The decision was rescinded the following day (Pomfret, *Washington Post*, November 23, 1996).

12.3. Economic Policies

The Croatian economy, second only to Slovenia in GDP and other indicators in pre-1991 Yugoslavia, deteriorated after the Serbian attack. Its chief sources of foreign currency, tourism and shipbuilding, have suffered. The government's attempts to privatize the Croatian economy have been frustrated. Its leaders have resisted pressure to pursue an inflationary monetary policy, but the tight policy adopted instead has resulted in currency supply problems. Economic reforms had taken place in Croatia during the 1980s as a response to the recessions experienced by the EE communist countries. In 1988 the so-called Markovic reforms introduced extensive price, import

Table 12.1 Economic, Social Indicators for the Danube River Basin Transition Countries, 1990 - 1995, Croatia

INDICATOR		1990	1991	1992	1993	1994	1995
Population, millions		4.8	4.8	4.8	4.8	4.8	NA
Population growth rate %		0.6	NA	0.1	0.1	0.1	
Infant mortality rate per 1000 live births		22.0	21.0	11.6	9.0	8.7	8.4
Life expectancy	(M)	70.0	70.0	NA	69.7	70.1	70.6
	(F)	76.0	76.0	NA	76.9	77.3	77.7
Literacy rate %		NA	97.0	97.0	97.0	97.0	97.0
Overall Economic Conditions		**1990**	**1991**	**1992**	**1993**	**1994**	**1995**
GDP in $ bn		NA	26.3	21.8	19.4	18.5	20.0
Real GDP growth %		-8.5	-20.9	-9.7	-3.7	0.8	2.0
GDP per capita in $		NA	5,600	4,500	NA	2,640	NA
Industry % of GDP		30.4	29.7	26.9	26.8	NA	NA
Agriculture % of GDP		10.6	10.2	13.2	12.6	NA	NA
Unemployment rate (% labor force, end year)		8.0	14.0	12.9	12.8	12.6	12.5
Retail price inflation (av) %		609.0	123.0	665.0	1,517	98.0	2.0
Exchange rate (CRD:$; HRK:$)		11.2	18.7	258.0	3,577	6.0	5.5
General govt balance, % of GDP		NA	-5.0	-4.0	-1.0	1.7	-0.4
Trade balance in $ m		-1,168	-536.0	137.0	-763.0	-969.0	-2,000
Current account $ m		-621.0	-589.0	329.0	571.0	625.0	-1,500
External debt in $ bn		NA	NA	NA	2.6	2.9	NA

and foreign exchange liberalization. While the attempts to initiate market reform were largely unsuccessful the effort helped in some ways to prepare for the transition period that was to come. The "social ownership" concept of the former Yugoslavia had already provided a degree of self-management to Croatian enterprises (Bicanic, 1994, pp. 626 ff).

12.3.1. REFORM PROGRAM

A comprehensive program of reform was adopted in the second half of 1991, leading to the creation of a central banking system, the introduction of temporary currency, and a national budget. The Law on the Transformation of Enterprises with Social Capital, passed in April 1991, is administered by the privatization fund and requires the conversion of almost all "socially owned" enterprises into joint stock companies. In October 1993 the parliament adopted a comprehensive macro economic stabilization programme. By January 1994 a number of reform tax laws had been passed (EBRD, 1994, p. 20).

The private sector accounts for 30% to 40% of Croatia's GDP. "Following two failed efforts, the third attempt to develop a coherent, macroeconomic stabilization policy, begun in October 1993 has had unexpected success and lowered inflation rates from a hyper-inflationary, 30 percent monthly level in September 1993 to negative rates for the first five months of 1994 (Bicanic, 1994, p. 622)." There has been a massive growth of new companies, from 55,000 in 1992 to 96,000 in 1994. This partial success story has taken place against a background of armed rebellion within its own country and caring for hundreds of thousands of refugees who needed shelter and food.

Entire geographical areas supplying agricultural products, among other commodities, have been cut off by the armed conflict. Energy sources are not operating. In 1994 the total direct cost of the war had been estimated to be three times the current GNP or US$22.5 billion. The government has laid the groundwork for economic transformation but its implementation awaits a firm peace which must be followed by the restoration of the infrastructure destroyed by the war.

12.4. Environmental Issues and Policies

Despite the hardships inflicted by conflicts within their country and in the neighboring republic of Bosnia-Herzegovina, the Croatian government joined the Danube Program in 1993 and continues to support for participation in its programs. During hostilities, government experts developed ways to protect residents from "ecocide" or "eco-terrorism" as destructive acts against the environment have come to be called. Among other steps taken, according to Zeljko Buksa, a writer on the staff of the largest daily newspaper in Zagreb, Vjesnik, reserves of toxic chemicals in warehouses were reduced to prevent dangerous gas emissions (Buksa, Danube Watch, 1:3, p. 4).

Oil refineries at Sisak and Basanski Brod which were in close proximity to the Sava River received several attacks during the Serb invasion. The Sava has been polluted by an estimated 12,000 tons of oil and oil derivatives since hostilities began. At Belgrade levels of minerals in the Danube were found to be 175 times the permitted maximum level. The historic city of Vukovar, which was almost totally destroyed, lost valuable natural areas from the toxic residue of the 300 tanks brought into the city.

12.4.1. MAJOR DANUBE ENVIRONMENTAL ISSUES

Croatia has a population of 4.8 million, 25% of whom reside in its capital, Zagreb, by far its largest city. The Danube drains an area of some 34,760 sq. kms., out of a total surface area of 56,500 sq. kms. The population of its drainage area was 3.2 million in 1991. Deaths and migrations as a result of the war have no doubt altered this figure. Most of the available data have not been updated since 1990.

Water Resources

The Danube catchment area includes two of the river's largest tributaries, the Sava and the Drava. Another river, the Mura, flows into the country and joins the Drava in the northern section of the country. The Sava is shared with Slovenia and forms a border of

312 kms. with Bosnia-Herzegovina before joining the Danube in Serbia. The Drava flows through Austria, Slovenia, Hungary, and Croatia, while the Mura is shared with Austria and Slovenia. The Danube catchment area measures 34,760 sq. kms; the Sava, 25,100 sq. kms. and the Drava 9,660 sq. kms.

Slightly more than fifty per cent of the Croatian population use the water supply system, while the rest draw water from personal wells. Public water supply systems are fed primarily by ground water and have been established in the more populated areas and industrial centers. A long-term plan anticipates that 90% of residents will be supplied by water supply systems by 2020. According to present forecasts, water resources will be sufficient for such a major increase in water supply facilities (IUCN, 1994, p. 117).

Water Quality

In the Croatian part of the Danube Basin about 30% of the population is connected to sewerage treatment facilities. Domestic waste of the rest of the population is contained in septic tanks or discharged directly into surface water. Cities and industrial plants have been discharging waste water into watercourses without or with minimum treatment.

Other Resources

The confluences of the Mura and Drava as well as the Drava and the Danube form valuable floodplains with a very rich fauna and flora. There are few data about Croatia's wetlands, but it was clear "even by 1990 that several important wetlands were threatened by planned water management alterations, for example engineering schemes at the mouth of the Mura (UNDP, 1994 34 ff)." The Lonjske Polje bird area, downstream from Sisak on the Sava, a unique collection of three ornithological reserves listed under the Ramsar Convention, is threatened by a petrochemical plant. The Strategic Action Plan identifies four sites for biodiversity restoration, including one war-damaged floodplain forest of 50,000 sq. kms. which requires biodiversity and war damage surveys, protection from pollution sources, especially agriculture, and detailed plans for restoration and maintenance of facilities at an estimated cost of MECU 10.

A number of dams and reservoirs (48 reservoirs of varying sizes on the Danube, Drava and Sava, with seven hydro-power plants) affect the natural sediment transport regime, increasing downstream erosion and retaining sediment behind the structures. Major industrial plants impact Croatia's environment (SO^2, particularate matter, among other emissions). Data are not available on possible high pollutant concentrations in the soil resulting from the use of fertilizers and herbicides but these are suspected. The oil pipeline connecting Serbia, Hungary and the Adriatic to the Sisak refinery has not been adequately protected and thus poses a permanent threat to soils, aquifers and surface waters.

Pollution Sources

The release of pollution is predominantly through point sources from heavy industrial plants located alongside the major rivers without adequate pollution treatment technology. Croatia is relatively industrialized and the Danube river basin is the center of most of the activity. The Zagreb-Sisak region includes chemical and pharmaceutical plants, paper and pulp mills, food processing factories and iron and steel works, in addition to the afore-mentioned oil refinery in Sisak. The largest food processing plant in the country is at Vrbovec and discharges untreated wastewaters with an organic load equivalent to a population of 185,000.

Significant diffuse oil leakage occurs from oil fields between the Drava and the Sava Rivers and from the pipelines which emanate from them. Further diffuse pollution comes from the effects of transport, since Croatia is an important crossroads and gateway both to the sea and to southern Europe. For example, in 1991 approximately 30,000 vehicles a day were using the highway between Zagreb and Belgrade, of which a quarter were heavy goods trucks (IUCN, 1994, p. 119).

Major pollution problems exist in all Croatian municipalities since few have sewage treatment plants. Wastewater from them is discharged untreated into the surface water network. There is no proper management or treatment of municipal solid waste which in 1991 totaled 800,000 tons. Zagreb alone generates about half of the Croatian contribution to the pollution load of the Sava.

No data were available on pollution from irrigation although it is likely to be insignificant because of a high rainfall. In the former Yugoslavia small private farms were maintained alongside larger state farms, and these different systems of production were characterized by significant differences in fertilizer and other input application. Pollution from agriculture is a source of concern since waste from pig farms goes untreated into streams. A high level of untreated solid waste disposal is also a problem.

Health and Hot Spots

Health studies have shown the presence of methemoglobinemia in infants as well as gastric cancer in Croatia, particularly in rural regions where well water is used (UNDP, 1994, p. 27). Extensive biological and chemical contamination in drinking water samples was reported in a 1992 government report. Lack of protection of the water supply in rural areas is widespread. The Strategic Action Plan recommends that 22 polluted sites receive treatment, with eleven of them identified as high priority. Of the 22 sites, ten require waste water treatment plant modernization or construction. Of these seven are high priority including three which must have war damage repaired. Modernization of industrial and other enterprises are among the remainder.

12.4.2. GOVERNMENT POLICIES

The constitution adopted by Croatia in 1989 gives the right to a healthy environment to all individuals and makes its protection a common responsibility. Natural resources are under the special protection of the government and must be used in compliance with the law. A declaration on environmental protection was adopted by the parliament in 1992 and general environment protection was provided by an environmental law enacted in November 1994. This act provides a framework for additional legislation, establishing basic goals and principles for environmental protection such as environmental impact assessments, the use of environmentally sound products, monitoring and information systems, methods of financing cleanup and the polluter pays principle (Mastrovic, 1995).

Legislation now in force remains fragmented, made up of over 400 regulations, which has been criticized for providing only marginal enforcement. The legislative basis of environmental protection is characterized by fragmentation, a lack of equilibrium in solving specific issues, and of regulatory co-ordination. There are major gaps in regulator provisions, many of which are inappropriate (Mastrovic, 1995, p. 37).

The Water Act of December 27, 1995 provides for improved reporting on water management conditions, including inventories of surface and ground water conditions. It also supports the construction of water management projects and water supply facilities. The act regulates water use and protects all surface and ground water. It also provides for user fees as the source of revenue for financing water protection. All legal persons and households must pay for discharging polluted water, following a complex formula. Water quality tests began in the 1930s on the Drava near Zagreb; today regular monitoring takes place at 60 stations. Key Institutions

A reorganization of government structure in 1994 made fundamental changes in the administration of environmental policy. All environmental activities of the Ministry of Civil Engineering and Environmental Protection were transferred to a new State Directorate for the Environment. The Ministry was renamed the Ministry of Spatial Planning, Civil Engineering and Housing. State Directorates for Cultural Heritage and Nature Protection and for Geodesy were also created. The State Directorate for Environment coordinates environmental policy and supervises compliance with legislation through the Environmental Inspectorate which was established in 1993. "Despite traditionally good cooperation between the State Directorate and other government agencies in charge of water management, nature protection and sanitation, there is still no ongoing cooperation between governmental agencies to create an environmental protection strategy and policy (Mastrovic, 1995, p. 36)."

12.4.3. POLICY OVERVIEW

Priorities have been established in a number of areas which may facilitate more effective environmental policies and their coordination. The State Directorate has an information program which will provide access to data about pollution levels, the state of the environment, international sources, among other aspects of the environment. A move toward enacting laws to implement the framework legislation is under way. Basic legislation which enforces standards for all media is fragmented and needs to be unified.

Public participation is encouraged and opportunities have been offered to nongovernmental organizations to receive information about policies and their results. These efforts are scattered. NGOs urged that the government create a cabinet-level ministry for the environment but this suggestion was rejected. Failure to achieve transition to a market economy and a debilitating budget crisis because of the war have set a very low priority for investment in waste treatment projects.

CHAPTER 13

Czech Republic

13.1. The Breakup

Political and economic change began in the Czech and Slovak Federal Republic (CSFR) in the years immediately following its rejection of communism in 1989 and before the agreement to separate into two states took effect on January 1, 1993. Overt opposition to the communist government began with student protests in November 1989. Pressure from opposition leaders, through an umbrella organization called the Civic Forum, forced the government to resign later that month. Playwright Vaclav Havel was chosen as president of the republic late in December. The new constitution and other reform steps taken over the next two years at the initiative of the CSFR were adopted by the two countries after their separation.

When elections in June 1992 brought Vladimir Meciar to power as premier of the Slovak Republic, both he and the newly elected premier of the Czech Republic (CR), Vaclav Klaus, favored the end of the federation. Havel stepped down from the presidency in July, while Klaus and Meciar completed a plan to carry out the agreement to separate by the end of the year. Vaclav Havel was elected president of the Czech Republic by parliament on January 26, 1993.

13.2. Western Orientation

CR foreign policy goals favor integration into the western community and cooperative regional relations. A ten-year trade and economic cooperation agreement between the CSFR and the EU was signed in 1990. A formal Europe Agreement reached in December 1991 was not ratified because of the pending separation of the two states. Separate agreements with the EU were signed with each country and ratified by the European Parliament in October 1993. Both receive assistance from the Phare program. The CR was the first former communist country in the Danube Basin to join the OECD

Application for EU membership was submitted in January 1996. There have been troubling questions regarding reparations owed by Germany to Nazi victims which have yet to be completely resolved. In 1994 the CR began to compensate these victims. Sudeten German refugee organizations threatened a stand against the integration of the CR with the EU if Prague did not fulfil their demands. The CR is a non-permanent member of the UN Security Council. It is a NATO partnership country; members of its military force are in Croatia with the UN forces.

13.3. Political Reform

13.3.1. CONSTITUTIONAL AND INSTITUTIONAL CHANGES

Most features of the constitution adopted by the CSFR in 1991 became part of a new CR Constitution. A Charter of Fundamental Rights and Freedoms obligates the government to protect a number of rights for citizens including the right to a healthy environment. A president is elected by Parliament for a five-year term with certain decisions subject to the approval of the Prime Minister. A bicameral parliament with special powers for the upper chamber has also been established. The upper chamber concept was not well received initially by all members of the parliament. It was voted in 1996 and the first elections for the 80-body Senate held in November 1996. All elections are free and based on direct universal suffrage. The role of the judicial system is essentially to defend the constitution. National administrative structures have been somewhat modified but the large number of counties, 72 in all, is considered by some to hamper effective elections and other political processes.

13.3.2. POLITICAL PARTIES--PARLIAMENTARY ELECTIONS

Prime Minister Vaclav Klaus is credited with understanding the need to seek and maintain strong political support for economic reform. Voucher privatization was widely popular and helped to provide Klaus and a coalition headed by his Civic Democratic Party with a victory in the 1992 parliamentary elections. In the parliamentary elections of June 1996, however, Klaus' ruling coalition failed to win a majority in the House of Deputies. The coalition won 99 out of 200 seats which meant that Klaus, to avoid a vote of no-confidence, would have to keep the support of two opposition parties, the Christian Democratic Union and the Social Democratic Party. A new minority government was installed in July, after the Social Democrats walked out, enabling the coalition deputies to pass a vote of confidence. The vote in the House was 98 to 40 in favor of the government. Agreement among the parties retains foreign policy priorities including full membership in NATO and the EU, and further development of the Central European Free Trade Agreement (CEFTA).

Two parties polled more than a third of the vote, the Christian Democratic Union, and the left-leaning Social Democratic Party which quadrupled its share of the vote to gain the second largest number of seats. The Civic Democratic Party is said to have become complacent and failed to appeal to the substantial part of the electorate which does not believe it is getting a fair share of the economic gains of recent years (*The Economist,* June 8 1996, 52-53). A number of issues including widespread evidence of corruption in public and private financial circles and failures in the health service have contributed to a shift in voter support away from the Klaus coalition and increased questions about the way the leader has exerted control over the coalition. Klaus continues as prime minister but there is general agreement that his leadership faces an uncertain future. The CR is still the only east European country where the former communist party supporters have no role in government.

Elections for the 81-member Senate were held in November 1996. Klaus's Party took 32 seats. Altogether the premier's coalition of three parties took 52 seats out of 81. The turnout in the electoral districts was low, 31% compared with 76% in the June 1996 elections. One-third of the Senate will be elected every two years; senators serve a six-year term.

13.3.3. THE CIVIL SOCIETY

While President Havel has argued for a healthy civil society, Klaus has refused to endorse the concept saying that it would introduce "third ways" into the building of a democratic society (Pehe, April 5, 1995). His reluctance to support or encourage activities of nongovernmental organizations may stem from his belief that a centralized system of government will be more efficient during the CR's period of rapid economic transformation. There are a large number of parties, 57 altogether, and 29 organizations characterized as movements. A very small number reach threshold limits for electing delegates to the parliament.

The CR has television channels and broadcasting systems financed from the national budget; freedom of the press and information is protected by the constitution. Despite the prime minister's failure to support a civil society role in government, the number of interest groups seeking access to government decision-making has increased considerably since the change in governments in 1989-1990. There are over 3,000 foundations and 22,000 clubs. Civic groups, trade unions, and employer associations are among the economic and agricultural organizations which have grown in strength. Umbrella organizations of such groups adds significance to their political presence.

13.4. Economic Policies

The CSFR inherited from the period of communism a low foreign and state indebtedness as well as a relatively small budget deficit and low inflation rate. It also had an almost completely state-owned economy, with only 2% of all assets owned by the private sector and only 1.2% of total employment part of that sector. It was "perhaps the most centrally planned and controlled state in the region, industrialized with an emphasis on military and heavy industry, and collectivized in agriculture (Kaiser, 1994, p. 507)." As in other countries in the eastern bloc, productive capacity was located and developed according to the perceived needs of the Soviet Union and CMEA. The overemphasis on heavy industry and weaponry was especially disadvantageous in the Slovak Republic.

13.4.1. MEASURING CHANGE

While the collapse of traditional markets has had a more severe impact in the Slovak Republic (SR) than elsewhere in Central and Eastern Europe, the CR has fared better and has been on a faster track to privatization. Its economic reform policies maintained

Table 13.1 Economic, Social Indicators for the Danube River Basin Transition Countries, 1990 - 1995, Czech Republic

INDICATOR	1990	1991	1992	1993	1994	1995
Social Indicators						
Population, millions	10.4	10.3	10.3	10.3	10.3	10.3
Population growth rate %	0.3	0.3	0.2	0.2	0.2	0.3
Infant mortality rate per 1000 live births	11.0	11.0	9.9	8.5	9.3	8.9
Life expectancy (M)	69.0	69.0	68.0	68.9	69.4	69.9
(F)	76.0	77.0	76.0	76.6	77.0	77.4
Literacy rate %	NA	NA	NA	NA	NA	99
Economic Indicators						
Overall Economic Conditions	1990	1991	1992	1993	1994	1995
GDP in $ bn	NA	NA	28.4	31.2	36	45.6
Real GDP growth %	-1.6	-14.2	-6.4	-0.9	2.6	4.6
GDP per capita in $	NA	6,900	7,300	7,200	7,350	NA
Industry & construction % of GDP	NA	NA	45	39.8	39.3	NA
Agricultural % of GDP	8.4	6	5.7	6.2	5.5	NA
Unemployment rate (end period)	0.7	4.5	2.6	3.5	3.2	NA
Consumer price inflation %	10.8	56.6	11.1	20.8	10	8.9
Exchange rate (av) Kcs/Kc:$	17.95	29.5	28.3	29.2	28.8	26.3
General govt balance, % of GDP	0.1	-2	-3.3	1.4	1	0
Trade balance in $bn (excluding trade with Slovakia)	-0.8	-0.4	-1.9	-0.2	-0.7	NA
Current account $ m	-949	328	-140	682	-81	-1350
External debt in $ bn	7.6	9.1	3.8	8.6	8.7	NA

micro-economic stability while keeping inflation at a 10% level, lower than any other post-communist economy in transition. The state budget has been balanced (*Financial Times*, 2 March 1994).

By late 1994 the Czech Republic was identified, along with Poland and Hungary, as being on the fast track toward privatization and economic improvements (*International Herald Tribune*, October 8-9, 1994, p. 1,5). Half of the work force was employed in private business. Scores of small manufacturing plants and private banks had opened. The International Herald Tribune of June 4, 1996 reported that Klaus's program had made the CR a model "of stability and free-market reforms among Europe's former Communist countries." It was pointed out that inflation had been kept under control, unemployment was stable at about 3% and "mass privatization had created a country of shareholders who still support reform (Green, *International Herald Tribune*, June 4, 1996)." The transformation into a market economy is generally considered to be nearly complete. "The share of public administration is reduced to the level usual in Western countries, the decisions of individual economic agents constitute the driving force of economic activities (Handl, 1995, p. 55)."

Competition from western European consumer products has resulted in decreased sales of domestic items. The CR has effected extensive price liberalization, devaluation of its currency, liberalization of foreign trade, and tight money supply control. It has restructured the state budget, emphasizing reductions in subsidies and increases in the prices of public utilities. State-owned property has been privatized and the tax system adapted to a market economy.

13.4.2. PROBLEMS

Problems have arisen, however, in connection with the banking system. In September 1996, the governor of the Czech National Bank took over the country's fifth largest bank, Agrobanka, following the arrest of one of its top executives who had been implicated by police in the collapse of another bank the previous month. In the latter case, the Kreditini Banka had collapsed with 12 billion koruny ($452.5 million) in losses. It was the eleventh Czech bank to fail in the last eleven years (Green, *International Herald Tribune*, September 21-22, 1996 11). Mr. Klaus acceded to the demands of the Social Democrats for a parliamentary investigation into bank failures.

The division of farm lands into small privately held, inefficient units which has taken place elsewhere in eastern Europe has been avoided in the CR. While the number of land owners is relatively high, cooperative successors to state owned farm collectives have been given government support so that about 2,880 cooperatives hold 3/4ths of the cultivated land. The sector employs five to seven per cent of the workforce. There are some complaints about the level of productivity and low profit margins.

13.4.3. SOCIAL INDICATORS

There has been a decrease in real wages of about 20% during the first transition years. There is a general public perception that there is an increase in poverty, problems of health, and the severe housing shortage inherited from pre-transition years. The quality of life, environmental conditions, health, and education are deteriorating. Racism has become a problem as Czech "skinheads" attack Roma. Kaiser notes that despite the positive course of the CR economy, "income disparities are widening, and people on fixed incomes are at risk if the social safety net is not secured (Kaiser, 1994, p. 510)."

The CR has received MECU 230 million ($299 million) from the Phare program from 1991-1995. In accordance with revised procedures (Chapter 25) project funds are directed toward the republic's preparation for EU membership. About one-third of the assistance funds have gone for infrastructure improvements (Jennewein, 1996, p. 71).

13.5. Environmental Issues and Policies

13.5.1. MAJOR ISSUES

Air and water pollution levels continue to affect the health and well-being of both rural and urban populations in the CR. Air pollution has had an especially severe impact in all industrial areas but especially in the mining districts of Northern Bohemia. Here and elsewhere in the CR emissions from stationary sources such as heat and power plants and metal and chemical processing units, among others, combined with automobiles and other forms of transport continue to create serious national and international problems.

Water Supply

Absence of water supply from mountain snow and dependence on precipitation make the country vulnerable to increased pollutant levels in surface and ground water supplies. Pollutants from human and industrial wastes are disposed of in rivers and streams with minimal or no treatment. Eutrophication affects the water supply as does damage from elevated salt levels, the result of poor soil management and other factors. Residential and industrial complexes, for the most part located along rivers, have become clusters of air and water pollution which affect not only the rivers themselves but neighboring agricultural development.

The CR, while not a riparian country, is served by the Morava River, one of the most significant Danube tributaries. Its total area of approximately 27,000 km^2 is primarily in the CR (84%) while 8.4% is in the SR. The section in the CR is a relatively densely populated area, averaging 1,230 inhabitants per km^2 and a total population of 2.735 million. The Morava basin varies greatly in terms of relief, geology and climate, ranging from the warm parts of the south, to the cold, mountainous regions in the north. The heavily polluted tributaries of the Morava continue to be a source of concern to environmental groups and to its general population.

Ground water supplies about 55% of the drinking water supply. Industry, particularly power generation, is the biggest consumer of surface water; its consumption accounts for 47% of total water consumption. The thermal (steam) power plant at Hodonin and the nuclear plant at Dukovany use vast amounts of water in comparison with all other consumers (IUCN, 1994, p. 122).

Water Quality

Surface water quality has been monitored in both republics since the early 1960s (Roneak, 1996, passim). The Czechoslovak Standard 'Classification of Surface Water Quality' provides for five classes of surface water, which range from 'very clean', or suitable for all purposes, to 'very heavily polluted', which is inadequate for most uses. The streams and rivers in the basin are heavily polluted. Almost 50% of the length of the largest tributaries of the Morava are categorized as heavily or very heavily polluted with respect to oxygen levels; 85% judged to be as heavily polluted with respect to chemicals. There is a high level of nitrates and phosphorus. The worst pollution from biological and microbiological factors are found downstream of the towns of Brno, Slavkov and Vyskov.

Use of agriculture chemicals, as it does elsewhere, threatens ground water resources. Waste chemicals and oil and petroleum produced by industry are also serious pollution problems. An estimated 60% of drinking water does not comply with legal standards and must receive series of treatments before it can be can be used.

Although water quality monitoring in the CR is reported to have taken place over several decades, the standards adopted and choice of parameters have been challenged as well as enforcement policies. Country experts believe that improved and more extensive monitoring is necessary. At a NATO Advanced Research Workshop held in May 1996, Jan Kubat, a Czech Republic specialist, said that market principles for data exchange had been introduced since the state budget does not fully cover the costs of this activity. Thus even exchange of data between institutes may carry a charge. Such practices may not support adequate monitoring; in any case there is a need to clarify its rules (Murphy, 1996).

Other Resources

Pollution has affected biodiversity in the Morava Basin. Some freshwater ecosystems are protected. In the lower reaches periodically flooded pools receive a great deal of chemical pollution; as they are filled in they are a serious threat to species habitats. The Danube Program Strategic Action Plan recommends eight projects to increase resource protection in the Morava-Dye area. The most extensive project would increase protection to 5,122 hectares in the Morava River floodplain to add to flood control, river biodiversity, forestry preserves and other uses (EPDRB, 1995, p. 26). Four proposed projects would help to remove nutrients, all of them in the Morava-Dyje basin.

The IUCN reports that there are currently no Czech standards for river sediments but a national study which sampled streams and reservoirs and compared results with Netherlands standards found that bottom sediments are not heavily contaminated by heavy metals. Polluted sediment layers were found to be situated mostly in the lower reaches of the river, off the main streamline of the watercourse.

Monitoring of a number of toxic air emissions, such as volatile organic substances, ozone and heavy metals, is increasing with help from the Phare program. Systematic monitoring and an evaluation system for soil quality is being developed. Monitoring in certain localities has revealed that the soil is contaminated by pesticide residues, polychlorinated biphenyls (PCBs) and heavy metals.

Pollution Sources and Treatment

The data available in the early 1990s showed that concentrations of pollutants from older food processing companies was high and that other types of industry were discharging a variety of toxic chemicals into surface waters. Currently, more detailed monitoring is being carried out and new standards are being set which will oblige polluters to monitor and treat their discharged wastes.

Most waste water treatment plants are currently in need of or undergoing reconstruction. All municipalities with over 50,000 inhabitants have at least minimal waste water treatment facilities (i.e. activated sludge process), but some smaller settlements lack the financial resources to build any such facilities. Although most settlements have a sewerage system, many systems are now outdated and not in good working order. Many of the plants are overloaded and none have the technical facility to decrease the nutrient content of waste water. Overloading can result in the discharge of untreated water, activated sludge run-off or leakage. The solution to the problem requires significant financial investment and improved coordination of the central administration of waste water treatment management. The overall direction of the program is reported to have poorly conceived and maintained (IUCN, 1994, pp. 26ff).

Chemical aids to agriculture pollute surface water in the Morava basin. Both fertilizers and pesticides contaminate water supplies by the processes of leaching and run-off (i.e non-point source pollution). The IUCN report found that major changes in economic structure are affecting agricultural patterns, and some environmental improvement is expected as a result. The manure produced by large-scale livestock farming is a major source of nitrogen pollution, particularly when it is produced as liquid manure, which more readily contaminates water supplies. Information on the location of animal farms is not readily available. The use of chemical fertilizers and pesticides has increased nitrate concentrations.

A number of priority landfill sites in the Morava Basin have been identified as potentially dangerous. Most sites are the result of poorly managed waste dumping primarily by industries. Treatment, including the sealing off of sites so that they do not pollute ground or surface water, has been planned but is not yet considered to be effective. The IUCN report found that water-related structures in the Morava basin are old and in need of reconstruction. Some reservoirs, first designed and built for protection against flooding, are used to store drinking water supplies, and also for irrigation, industry and recreation purposes. Certain flood control operations have caused changes in the natural environment which require a systematic study of their effect on biodiversity.

Health and Hot Spots

The Strategic Action Plan recommends ten projects to improve waste treatment in the CR. Five are considered to be high priority. Four are for waste water treatment plants, of which two are high priority. With only one exception pre-investment studies have accurately estimated the costs of the proposed projects. The municipal WWTP at Brno, a major city on the Morava, would modernize the sewerage network and reconstruct the system so as to provide drainage and treatment for municipal and industrial waste waters as well as the disposal of contaminated sludge. The estimated cost is US$39.75 million. At another major city, Prerov on the Morava, a reconstruction of chemical industrial plants and a sewerage network would reduce threats to ground water sources used for drinking water. The total estimate for the projects is US$15.87 million.

Both national and international goals include specific clean-up projects which are still in the pre-investment study stage. Project Morava will improve water quality in the Morava river basin, reducing pollution in the Danube, Project Dunaj focuses on water quality in the Danube and lower Morava; and waste management programs which focus on solid wastes disposal. Changes are being made to agricultural methods and there are also moves to encourage the development of 'clean' technologies and to reduce the consumption of materials and energy. The construction of waste water treatment plants is a priority measure, the expected cost of which is estimated at 520 million Kcs ($17 million). International support, in addition to that proposed by the Danube Program, include the Project Danube Morava, funded by the World Bank and a study of water quality in the Dyje catchment area, funded by Allplan Wien of Austria.

13.5.2. GOVERNMENT POLICIES

The Constitution adopted by the CR in December 1992 provided guarantees for a right to a favorable environment and information about its status, and the requirement that no one endanger the environment. Environmental problems received attention from 1990 to 1992 as "active participation of non-governmental leaders in the policy-making process and massive feedback from NGOs to the government" occurred (Regional Environmental Center, May 1995, p. 43). The Federal Act of 1992 defined the basic terms and principles of environmental protection as well as the legal obligations of individual persons and legal entities to protect the environment. It established general principles for sustainable development and providing accurate information about the state of the environment. It also provides for environmental impact assessments (EIAs), and environmental liability.

Public Support

A significant decline in public support for environmental measures took place in the CR in 1993 and was reflected in the policies of the new prime minister, Klaus. Election polling showed that economic issues and increasing crime were of greater concern than the state of the environment, which fell to sixth place as a priority concern. The Prime Minister affirmed his own lack of interest in aggressive measures to control the effects of widespread air and water pollution by likening the economy to a cake and environmental protection to the frosting on it: you had to have a good cake before you could apply the icing (Regional Environmental Center, May 1995, p. 43).

Response to EC Mandates

Many of the laws still in effect, adopted immediately after the 1989 revolution, have problems stemming from "the obscure formulation of responsibility at various levels of administration and the ensuing difficulties in controlling the implementation and observance of the legal norms [established] (Kocikova, 1996, p. 29)."

Changes in CR environmental policies have occurred recently in response to two mandates. First, the European Agreement between the CR and EU which went into force on February 1, 1995 obligated the CR to harmonize existing and new legal provisions, technical standards, and relevant inspection procedures with those of the EU, including those which relate to the entire spectrum of environmental protection. Secondly, the CR government in response to the Lucerne ministerial conference is preparing an environmental action plan following a set of specific requirements.

The updated statement of national environmental policy adopted in August 1995 emphasized a number of principles relevant to EU environmental policies. These included the use of economic incentives to stimulate environmental protection including taxes, fees, and the like. The plan endorses an integrated approach to the environment encouraging realistic evaluations of environmental damage, international cooperation and the completion and adoption of a system of environmental laws. Many areas where legal norms are missing or not complete were given special attention. The plan provides a guide to future legislation and is intended to speed the satisfaction of EU requirements.

Water quality legislation is under revision. As in other media areas, Phare has provided partial funding for an analysis of legal norms for ground and surface water quality for drinking and other uses. An amendment to the 1973 Water Act was going through the legislative process in 1996.

Key Institutions

The Ministry of Environment administers environmental policies through various sub-units including the Czech Inspectorate for the Environment and the State Environmental Fund as well as specialist and scientific institutes. Other ministries have roles to play including the Ministries of Trade and Industry, Economy, Health, Agriculture, Transport and the State Nuclear Safety authority. The CR Ministry of Environment has prerogatives over air, water and soil quality, waste treatment, nature and landscape protection, forestry, hunting, fishing, and geological and mining surveys. The Ministry has under its supervision boards which manage river basins. It is also responsible for environmental impact assessments and for developing and releasing environmental data to the public. The ministry also supervises research agencies. These include the Hydrometeorological Institute which is responsible for monitoring and developing environmental data.

Other institutes have responsibilities for training programs in environmental impact assessment preparation, research related to water supply management and waste water treatment. The Ministry of Health regulates noise and toxic substance pollution and is responsible for general health and safety matters.

The Ministry of the Environment of the Czech Republic has a main office in Prague, and nine territorial departments. The Ministry is divided into departments which deal with the various elements of the environment and also with policy, legislation and the State fund for the environment.

Due to the rapid political and economic changes taking place in the Czech Republic, it is possible that further alterations will be made to the environmental policy structure in the future. The company Povodi Moravy, is entrusted with catchment and streams administration, and general water resources management and runoff control in the Morava River Basin.

13.5.3. POLICY OVERVIEW

According to an article in the Danube Watch, "Just the Cream on the Cake"?, by Ivan Rynda, former chairman of the CR Parliamentary Environmental Committee, (*Danube Watch*, 1:4, 4)17 most environmental legislation was passed during the first two years of the transition. Since the election of June 1992 only one such law has passed. Rynda faults the political leadership, in particular Prime Minister Vaclav Klaus who has said that protecting the environment is not essential to economic reform. The protection of nature is ignored and no attention paid to the protection of non-renewable resources.

Changes in the governmental system of the country and in its relations with other countries have caused significant changes in environmental policies. In past years, monitoring and evaluation of the environment were carried out but there was insufficient economic and technical support for follow-up measures. There was little attempt to resolve problems which were perceived as local issues although their serious wider impact was apparent. A more urgent need to make substantial improvements to the present poor state of the environment is generally perceived .

Current activities focus on more extensive and improved monitoring of hazardous substances and their effects; analysis and evaluation of the state of the environment according to new approaches enforced in developed countries which take account of sustainable development needs, and the development of new measures, including the application of advanced technologies and the monitoring and control of those measures to improve environmental conditions. The lack of financial resources with which to purchase new equipment and the latest technology is the main barrier to realizing these objectives. Their implementation await a more positive approach by parliament and prime minister.

CHAPTER 14

Hungary

14.1. Hungarian Reforms

Reform begun in Hungary in the mid-1960s has helped to advance its political and economic development during the early transition period. Encouraged by Gorbachev policies and driven by the pervasive recession of the 1980s, Hungary's leaders were ready by 1989 to transform the country into a multiparty parliamentary state. Independent parties were approved in that year; the beginning of a new era was confirmed by the retirement of Janos Kadar who had ruled the country since 1956. By September 1989 the Communist Party renounced Marxism and renamed itself the Hungarian Socialist Party. On October 23, the date of the start of the 1956 rebellion, the country was declared to be a free republic.

The present government is ruled by a parliamentary coalition of the Hungarian Socialist Party and the Alliance of Free Democrats with Gyula Horn serving as prime minister. Arpad Goncz was re-elected President on 19 June 1995 with a large parliamentary majority, to serve until the year 2000. The next parliamentary elections will be held in 1998.

14.1.1. WESTERN ORIENTATION

West European leaders have looked with favor on Hungary's economic and political progress during the early transition years and welcomed its strong western orientation. An EU Association Agreement was negotiated in December 1991 and has been in force since February 1994. The country is a member of the NATO Partnership for Peace program and the OECD. The 1995 meeting of the NATO General Assembly was held in Budapest in May, the first time it had been held in a non-NATO country.

Regional cooperation has been a foreign policy goal. At a meeting in January 1991 in Visegrad, an historic town on the Danube north of Budapest, trade and economic cooperation generally were agreed to by Hungary, Poland and the then CSFR. The four countries continue to be described as the Visegrad countries and have held additional meetings to implement the agreements reached in 1991.

An important item on the Hungarian foreign policy agenda has been the protection of large Hungarian minorities in Romania (two million), and in Slovakia (six hundred thousand), a legacy of the World War I Trianon Treaty. While satisfactory treatment has been accorded such minorities in Ukraine, Croatia and Slovenia, disagreements with

Slovakia and Romania continue. The Stability Pact agreed to in 1995 by Hungary and Slovakia which was intended to ensure fair treatment of the Hungarian-speaking minority has been ratified by both countries. Slovakia, however, included reservations when it ratified the agreement. Its leaders deny that the Slovak government has accepted the collective rights and autonomy of the Hungarian minority. Romania and Hungary signed a similar agreement in August 1996.

14.1.2. CONSTITUTIONAL AND INSTITUTIONAL CHANGES

A provisional constitution, essentially the same as one adopted in 1948, was endorsed after the 1989 change in governments. It was amended to some degree in the early period of the new government. After that the coalition in power, the Hungarian Democratic Forum, saw no need for a new document. When the Socialists and Free Democrats were elected in 1994 they kept a campaign promise and prepared a new constitution with a similar governmental structure. There is a one-chamber parliament with jurisdiction over the courts and with the power to elect the president. The new draft constitution was released to the public on Friday, March 29, 1996 and subjected to debates within parliament through 1996.

The Civil Society - Parties

In October 1989 the Hungarian Socialist Workers Party became the Hungarian Socialist Party (HSP) and voted to discard the name Hungarian People's Republic in favor of the Republic of Hungary. Despite other attempts at reform, the HSP was defeated when elections were held in March and April 1990.

A new democratic government led by the Hungarian Democratic Forum (HDF) and its allies, the Independent Smallholders' Party (ISHP) and the Christian Democratic People's Party (CDPP), took office in May. This right-center coalition gained 60% of the vote and 230 seats in the 386-member National Assembly. Led by Jozsef Antal of the HDF, it had a program emphasizing personal and political freedom, human rights, a social welfare safety net, and Hungary's membership in the EU. The HDF's most important rival parties include the Federation of Free Democrats (FFD), which favors even more rapid westernization, and the HSP, which hopes to preserve some aspects of socialism. In the 1990 elections the FFD won 23.6% of the vote and 91 seats, and the HSP, 8.5% and 33 seats.

In the May 1994 elections the opposition Hungarian Socialist Party (MSZP) and its ally, the Alliance of Free Democrats (SZDSZ), won 72% of the seats in parliament. The change in voter support appears to be less a popular wish to return to socialism than a desire to follow a more moderate approach to economic reform. The December 1994 local elections reaffirmed a center-left trend. The high rate of participation in the 1994 elections was essentially a support for the center; the extreme left and right parties received low percentages of the vote.

As part of the program for its four years in power, the government pledged that it would pursue transformation to a market economy, support parliamentary democracy, social justice, and integration into the EU. The modernization of Hungarian society and economy was to develop "a socio-economic model merging with an efficient market economy, based on the predominance of private ownership and economic competition, with social welfare reducing inequalities of opportunity (Pomazi, 1995, p. 51)."

Media

Newspapers represent a spectrum of political views and are generally similar in style and substance to those in the west. Continued state ownership of television and radio facilities has caused controversies which are not yet resolved. EU reviewers concluded that overall "...the Hungarian media are not less neutral and non-partisan in their every day information activities than the media in countries with longer democratic tradition and institutions (Inotai, 1995, p. 92)."

Interest Group Representation

The inherited political culture is authoritarian, "a civil society in which democracy is nurtured by leaders who emerge from, and are supported by the interest groups they represent, did not take hold immediately (Marer, 1994, p. 490)." Of special interest was the movement to stop construction of the Gabčikovo-Nagymaros dam project which enlisted a large number of individuals and groups in protest actions beginning in the mid-1980s (Fitzmaurice, 1996). While equated with a movement for greater democracy, its composite groups lacked democratic processes of their own and did not appear to have very many concerns about the fate of the environment elsewhere in the region.

14.2. Economic Policies

Economic reforms were in progress in Hungary for three decades before the official end of communism in 1989. Market-oriented initiatives were provided by the "new economic mechanism" which encouraged the efficiency of state-owned enterprises through profit incentives. Reforms had many positive effects. Management techniques improved and consumers became accustomed to a market system. An unfortunate legacy of this period of quasi-market reform, as Paul Marer points out, was that it accustomed people to semi-legal activities and pervasive corruption. "Stealing state property and evading personal taxes became morally acceptable, a hindrance in establishing a normal market economy (Marer, 1994, p. 484)." These practices continue to haunt the Hungarian economy. Nevertheless, as early as 1989 the modified communist government had advanced the country an estimated one-third of the way toward a market economy.

14.2.1. MEASURING CHANGE

Despite its history of market-oriented reform Hungary faced many of the severe economic problems experienced elsewhere in eastern Europe during the early transition years. The first post-communist government "did not have a clearly articulated economic strategy, nor did the government attempt to build a broad political consensus

Table 14.1 Economic. Social Indicators for the Danube River Basin Transition Countries, 1990 - 1995, Hungary

INDICATOR		1990	1991	1992	1993	1994	1995
Social Indicators							
Population millions		10.4	10.3	10.3	10.3	10.3	10.2
Population growth rate %		-0.1	-0.1	-0.1	-0.7	-0.03	0.02
Infant mortality rate							
per 1000 live births		14.8	15.6	14.1	13.3	11.6	11.9
Life expectancy	(M)	67.0	68.0	66.0	66.8	67.4	67.9
	(F)	75.0	76.0	75.0	75.1	75.6	76.1
Literacy rate %		99	99	99	99	99	99
Overall Economic Conditions		1990	1991	1992	1993	1994	1995
GDP in $ bn		NA	60.1	55.4	57	58.8	NA
Real GDP growth %		-3.5	-11.9	-3	-0.8	2.9	2
GDP per capita in $		NA	5,700	5,380	5,500	5,700	NA
Industry % of GDP		28.8	25.5	26.4	25.2	25.9	NA
Agriculture & forestry % GDP		9.6	8.6	7.3	6.4	6.6	NA
Unemployment rate							
(in % of labor force)		1.7	8.5	12.3	12.1	10.4	NA
Consumer price inflation %		28.9	35	23	22.5	18.8	28.2
Exchange rate (av) Ft:$		63.2	75.7	79.0	91.9	105.2	125.6
General government							
balance in % of GDP		0.5	-2.2	-5.6	-6.4	-8.2	NA
Trade balance in $ bn		0.3	0.2	0	-3.2	-3.6	NA
Current account $ bn		0.1	0.4	0.3	-4.3	-4.1	-2.4
External debt in $ bn		NA	22.7	23.5	24.7	27	NA

to help it design and implement an economic transformation and recovery program, even though such an act of statesmanship might have been desirable and politically feasible (Marer, 1994, p. 489)." There was no specific economic strategy in place and serious disagreement existed in the government over how to face budget deficits and other problems. Under Jozsef Antal, the prime minister from 1990 to 1993, legislative and institutional changes essential to a market economy were made. In any case, foreign investors funded Hungarian enterprises more than those of any of the other economies in transition.

Bokros Package

Hungary's problems follow from the failure of its governments to develop and implement economic strategies. Marer poses and attempts to answer the question: what economic strategy could have been pursued during 1990-1994? and matches optimal options with the actual decisions. While a risky undertaking for an international economist, his conclusions are enlightening and of value as Hungarian economic policy continues to unfold (Marer, 1994, pp. 491 ff). By the end of 1994 Hungary's foreign debt had increased to $31.8 billion and its current account deficit was at $3.9 billion. It consequently lost credibility in international financial markets and among foreign investors (Szilagyi, March 13, 1996). In order to be able to receive additional IMF support, a reduction in foreign indebtedness was required. Lajos Bokros was appointed as finance minister in 1994 and developed the "Bokros package" to produce the required reforms.

The stringent reforms included "eliminating macro-economic disequilibria" and removing Hungary's long-standing structural problems. The macro-economic component of the program was announced in March 1995, and includes an immediate reduction in the fiscal deficit, to be followed by additional reductions in subsequent years, devaluation of the forint, and a restrictive wage policy." The structural component included a "medium-term reduction in the size of the public sector, an ambitious program of enterprise privatization, and privatization of state-owned commercial banks (World Bank, 1995, pp. 11ff)."

In 1995 prime minister Horn revised the Bokros stabilization program and began a new program under the joint direction of Bokros and Gyorgy Suranyi, the president of the Hungarian National Bank. By the end of 1995 the current account deficit was lowered to $2.5 billion and the budget deficit to 6% of GDP. Hungary received a three-year standby loan of $300 million and was admitted to the OECD. The general unpopularity of the Bokros program among government parties and the general public caused Bokros to resign in February 1996. The austerity program continued, in modified form.

Agriculture

Hungary experienced two intensive farm collectivization drives in the 1950s and 1960s. These are generally regarded as a successful combination of mechanized crop production with farmer initiative. For ideological reasons the collectives were forcibly broken up during the early transition and replaced with family farms. As time went on these small sized units did not prove to be economically viable. Problems continue for Hungary for this reason and because of the protectionist agricultural policies of the EU. The share of Hungarian exports to the EU has dropped by 50% (from 25% of total exports to 13%) since 1990 (Inotai and Notzold, p. 105; see also Marer, 1994, p. 500-501).

14.2.3. SOCIAL INDICATORS

The modified support for socialism reflected in the current government's policies has promised to continue to provide a safety net welfare program while at the same time implementing the financial austerity required by western funding. Inflation is a serious problem for most Hungarians. Unemployment outside of Budapest is considerably higher than it is in the capitol. As an indication of the severity of social problems, public support for the new market system fell to 15% in 1995. Under the present tax system, which is one of the world's heaviest and most cumbersome, a sizable black market flourishes. It affects not only consumer sales but the construction industry and health services where tips and bribes are expected (Marsh, *Financial Times*, April 6 - 8, 1996).

14.3. Environmental Issues and Policies

14.3.1. ENVIRONMENTAL STATUS

A detailed and objective appraisal of Hungarian environmental conditions was provided in a 1990 "Yearbook" prepared by the Hungarian Academy of Sciences, the Ministry for Environment and Water Management, and the Hungarian Central Statistical Office. A basic finding: most of the changes of the post-World War II years which included rapid industrialization and urbanization, and increased agricultural production, took place without regard for "environmental consequences (Vukovich, 1990)." Ground and surface waters were heavily polluted, the latter from industrial and municipal waste and the former from mining activities. Only 52% of the population of 10.3 million were connected to sewerage systems. Another important problem inherited from past neglect is insufficient capacity for wastewater treatment; only 28% of total waste water receives adequate treatment.

Environmental protection policies, at least partly the product of Hungary's dependence on outside sources to meet steadily increasing water resource needs, had their roots in the pre-transition era. Legislation in 1976 provided a framework for environmentally protective measures and for a ministry of environment established in 1987 which was to coordinate policy on environment protection and water management.

Water Resources

Ninety-six per cent of Hungary's surface water resources come from outside the country. As the IUCN report points out, the quality of its surface water is therefore continuously and strongly influenced by events and trends in upstream countries. Hungary shares borders with seven other Danube Basin countries: Austria, Slovakia, the Ukraine, Romania, Serbia, Slovenia and Croatia. The environmental profiles of all of these countries, with the exception of Austria, are not reassuring with respect to the impact of pollution. Most of the ground water is bank-filtered, stored in alluvial formations along the Danube and Tisza Rivers. With respect to water quantity management, the most critical time for management is August, when the discharge of the main rivers is low and when simultaneous usage is highest. Most drinking water supply (90%) comes from subsurface resources. The drought of the last decade has seriously affected shallow groundwater and karstic system water resources.

Water Quality

Hungary has monitored water quality at sampling sites (250 in 1990) for a period of 25 years. Surface waters are classified as I, high quality needing little treatment for any use; II requires some form of treatment before use but is not detrimental to the system's ecological health, and III waters so polluted with industrial, municipal and agricultural waste that they cannot be restored by acceptable treatment methods. According to 1986-1990 figures 21% of the total river length was in class I, 61% in II, and 18% in III. Pollution sources include insufficient communal sewer systems and wastewater treatment facilities, partly treated industrial effluents, and the use of agro-chemicals (IUCN, 1994, p. 142). There has been little change in these figures pending the implementation of the National Environmental Action Plan and the possibility that Hungary will upgrade its water management program.

The river system entering Hungary is polluted from time to time by accidental discharges of industrial wastes such as oil and its products. As a consequence the surface water intake of the Budapest water works has to be closed down for periods of time. The frequency of such incidents will probably increase with traffic from the new Danube-Main waterway and anticipated restoration of shipping upon the final resolution of the Bosnian situation. The need for the establishment of an early warning system to protect users against pollution has been recognized and its development almost completed.

A large proportion of riparian forests have been lost due to modifications of the river system and the failure to replace forested ares. Impacts around the major cities and in woodland areas are particularly acute (Rakonczay, 1990, p. 127). The Strategic Action Plan identifies 13 areas for rehabilitation or improved management. In response to concerns about water resource protection, national parks are recommended for a 201 km section of the Upper Tisza and a section of the Drava basin (EPDRB, 1984, pp. 28-29).

There is as yet no classification system for the evaluation of sediment quality in Hungary. Substitute analytical methods (1988) measured heavy metal concentrations which significantly exceeded the limits for heavy metals in several water courses. Hungary's nationwide air-monitoring network reported in the late 1980s that 11% of the total territory, inhabited by 44% of the population, was classified as air polluted. The main sources are power plants, metallurgy and cement industries emitting sulphur dioxide, solid particles, carbon monoxide and nitrogen dioxide.

Greater Budapest and the Borsod region are by far the most polluted by sulphur and nitrogen dioxides, and the Borsod region suffers additionally from high dust loads. Costs in terms of human health and lost work hours have been estimated at 3.7 billion forints a year or .14 % of the Gross National Product and 13.3% of its health and social welfare budget (Varkonyi and Kiss, in Hinrichsen, 1990. p. 56).

Other Resources

Soil contamination is the result of poorly designed or illegal disposal sites, leakage from fuel storage tanks, chromium pollution of soil and groundwater in the vicinity of chemical plants, mercury pollution from an electrolysis plant, lead pollution from a heavy metal recovery factory, and general soil contamination around abandoned Soviet military bases. So far, more than 170 former military sites have been found to be loaded with hydrocarbons, heavy metals and other pollutants.

Pollution Sources

According to the IUCN report industrial waste water volume comes primarily from metallurgical enterprises (31.6%) and the chemical industry (27.9%). With light industries (13.7%), machine industry (11.1%), food industry (7.2%), power industry (4.3%), and mining (2.9%) making up the rest. Waste from the largest contributor, the metallurgy industry, is a particular problem since only 3% of is treated biologically or chemically. The heavily polluting chemical industry is concentrated along the Danube and Tisza Rivers. Polluted areas cover only one-tenth of the country but affect 40% of the population.

Municipal Discharges

Ninety two per cent of the population is served by communal water supplies and 52% are connected to the sewerage system. The majority of untreated domestic effluent is released from Budapest and downstream into the Danube and three sites located along the Tisza. Seventy-seven percent of sewage generated daily in communities passed into sewerage systems, but of this only 42% flowed into treatment plants. An estimated 20% of municipal discharges received adequate treatment (UNDP, 1993, p. 72).

Agriculture, Solid Waste

Since 1950 more intensive agriculture has been the occasion for the increasing use of chemical fertilizer. High nitrate concentration is a country-wide problem. The over-use and mis-use of fertilizers and pesticides are a serious environmental threat. Manure waste from large-scale livestock farms contributes to pollution as well.

Treatment of the 100 million tons of solid waste from industry and 22 million cubic meters of liquid and solid communal wastes every year is inadequate. While re-use and recycling of harmful waste is increasing the introduction of proper pre-treatment technologies is slow. Nearly 90 per cent of all municipal solid wastes collected in the country are dumped untreated into landfills. Hungary is already experiencing a critical shortage of suitable dump sites. The capacity of selective collection, recycling, and supplying devices is as yet underdeveloped in Hungary.

Engineering Structures

Extensive modification of the river to increase agricultural areas and to protect settlements from flooding began in the last century. As flooding accelerated, the impact of engineering works on river margins, floodplains and channel hydraulics had a special importance because of Hungary's dependence on bank infiltrated water for its water supply. Maintenance of water levels, limitation of dredging works and waste deposits near wells must be monitored and controlled.

The possible adverse effects on Hungarian water supply of the Gabcikovo-Nagymaros hydropower project has been the occasion for concern since its construction began in the 1980s. Diversion of the Danube has lowered flows and river branches and as a result riverine water bodies have dried out. Further impacts include a decrease in groundwater tables and a change in the flow direction with severe damage to the aquatic environment anticipated (IUCN, 1994, p.150).

Hot Spots and Health

A report on public health (Hertzman, 1995, p. 71) noted improvements in water quality in the early 1990s but stressed that this was due to an overall reduction of point source pollution. Non-point source pollution, particularly from the use of fertilizer and pesticides in agriculture, discharges from intensive livestock breeding, run-off from sealed urban and industrial surfaces, and seepage from land-fills and sludge deposits, remains a serious problem. Excessive nitrates in drinking water, has caused a relatively high number of cases of blue baby syndrome, a frequently fatal illness.

The Strategic Action Plan has identified 29 hot spot sites in Hungary; 16 are high priority. Seventeen of the hot spot projects are for waste water treatment plants; 14 are high priority. All but five of the recommended projects have been cost estimated, a high percentage for Action Plan targets, reflecting the optimism western investors have identified with Hungary and the Czech Republic (see above). By far the most ambitious project is the municipal waste water treatment plant for Budapest which would upgrade and enlarge the sewage treatment capacity of two existing plants and build two new plants for a cost of $540 million. This figure does not include anticipated sewer construction costs. Modernization of industrial and agricultural enterprises make up the remainder of the projects.

14.3.2. GOVERNMENT POLICIES

Hungary adopted framework environmental legislation in 1976. This act included some of the more advanced norms of western Europe. The pace of follow-up legislation was not maintained however. "...[T]he first years of the developing market economy were to promote a relatively anti-environmentalist policy, based on a misunderstanding of real values and the real costs of pollution (Erdey, 1995, p. 53)."

In 1994 the Ministry of Environment and Regional Policy produced an environment policy concept paper to guide a set of policy changes over a four-year period. Goals included most of the objectives identified in EU council directives such as the application of the polluter pays principle, a public information system, greater transparency in the adoption and execution of environmental regulations, among others. The assignment of short-term goals such as "elaborating and launching a detailed assessment program on the state of the Hungarian environment," estimated to take from one to six years, is useful but the willingness of the present government to allocate funding and enforcement authority to such goals has not yet been evident.

Implementation of new framework legislation enacted in 1995 is under way. Water quality issues are receiving special attention including higher penalties for river and lake, and sewage water pollution. Enforcement of pollution control laws depends on the emission and discharge licenses granted to enterprises and fines for failing to meet emission limit values. Modernization of the water legislation framework has received support from the EU Phare program but is not yet complete.

14.3.3. KEY INSTITUTIONS

The Ministry of Environmental and Regional Policy is responsible for providing central state control over environmental protection, regional development and the building sector. Its Environmental Protection Inspectorate (central office plus 12 district offices) is in charge of permitting, monitoring, enforcement and remedial action related to air and water pollution control, disposal of hazardous wastes, noise pollution and waste, noise and radiation management.

The State Public Health and Medical Officer Service is concerned with implementing legislation which protects public health, including the improvement of water quality. Various other ministries perform functions which impact on water protection matters. Local governments operate water and sewage treatment plants and water distribution. A number of special institutes have responsibilities, in particular VITUKI, an acronym which translates into English as the Research Center for Water Resources Development.

14.3.4. POLICY IN TRANSITION

A major overhaul of the system of environmental law follows current trends to restrict the state role to control and licensing of water use and discharge, while investment and financing of water collection, distribution, and treatment may be privatized. Water quality standards are to be harmonized with those of the European Community.

The Ministry for the Environment and Regional Policy has devised an action plan to deal with the current environmental issues. At present Hungary has neither the legal instruments nor the financial resources necessary to protect the environment effectively. Licensing is said to be required for any activity which might change the natural condition of water resources, air or soils. In practice, monitoring is not adequate and the system is ineffective. Achieving action plan goals will require much work and a large number of initiatives. On the matter of financing these initiatives, the effective utilization of foreign grants and loans and the reformed regulation of funding, in particular the Central Environment Protection Fund is vital.

CHAPTER 15

Romania

15.1. Timisoara Uprising and Aftermath

An uprising in December 1989 led to the assassination of Nicolae Ceasescu, Romania's dictator since 1965. Demonstrations against the Communist regime had taken place in several Romanian cities. On December 17 a crowd in Timisoara, protesting the treatment of Laszlo Tokes, a minister in opposition to the government, was fired upon, with many deaths. On the 21st a supposedly friendly, but arranged rally in Bucharest turned hostile and drove Ceasescu and his wife to the roof of Communist Party headquarters. From there they were taken away by helicopter. Ion Iliescu, with other leading Communists formed the Council for National Salvation and assumed control of the government. A military court ordered the execution of Ceasescu and his wife on December 25 (Carothers, 1996, p. 10; Perlez, *New York Times*, December 25, 1994).

The faces and names of those who held power until the election in November 1996 were the same as many who took part in the Ceascescu government. Those who have become wealthy in the change to a market economy are former communists. Ion Iliescu, Romania's president, was a Ceausescu supporter; Virgil Magureanu who headed the secret police held a similar post.

Decisive Elections

The results of the parliamentary and presidential elections of November 3, 1996 made it certain that changes in policy would take place. After beating the former communists in local elections in June, the Democratic Convention, a center-right alliance of groups opposed to Iliescu's leadership, won 30% of the seats in parliament, eight percentage points more than the former ruling Party of Social Democracy. A coalition was formed by the Convention with the Social Democratic Union, another reform-minded opposition group which came in third with 13% of the vote. Iliescu failed to get a majority in the presidential elections and was forced into a run-off with Emil Constantinescu, leader of the Democratic Convention, which the latter won on November 17, 1996.

Iliescu and a supportive parliament had failed to implement economic reform. Loyal to a constituency of farmers and supporters of the safety net provided by the former communist regime, his government had taken measures to prop up the local currency, the leu, by interfering with foreign currency exchange. As a result the International Monetary Fund held up loans the failing economy badly needed. The new government leaders support a speed-up of economic reforms by making private property ownership easier and by finally dismantling the large number of state-owned industries (Perlez, *New York Times*, November 5, 1996).

15.1.1. WESTERN ORIENTATION

Romanian foreign policy goals, reinforced by the new government leaders, include integration with the EU, and its security arm, the Western European Union, and NATO. It also seeks good relations with the west generally and with its neighbors. It is part of regional economic organizations such as the Black Sea Economic Cooperation Zone and belongs to a number of international organizations promoting regional and international goals for human rights, resource protection, and public health, among others. It has achieved democratic control of the military and supports a modern, western-style armed forces organization.

Romania began negotiations for a trade agreement with the EU in the 1980s and was the first COMECON country to do so. Negotiations were broken off in 1989 and not concluded until March 1991. The Europe Agreement required for entry into the EU was signed in 1993 and awaits ratification by the EU member states.

15.1.2. CONSTITUTIONAL AND INSTITUTIONAL CHANGES

Following the first post-war elections in May 1990 a new constitution was adopted in November 1991 and confirmed by a general referendum the next month. A strong presidency was established. Laws are introduced by the government or the parliament and have to be ratified by the president. The president, elected by popular vote and limited to two four-year terms, has the power to introduce amendments during the legislative process. The president may be impeached by parliament.

The constitution and additional legislation provides the legal basis for the separation of powers, a return to the rule of law, and an independent judiciary. Judges are, according to the constitution, independent and appointed for life. The constitution has been declared by the Council of Europe to be in accordance with European Human Rights Convention with respect to the protection of national minorities and in other ways.

Ion Iliescu, head of the National Front, was elected president in 1990 with 85% of the vote; his party won the chamber of deputies with 66%, and the upper chamber with 67%. Opposition parties coalesced under the Democratic Convention. While many local elections were won by the opposition, in 1992 Iliescu's National Front again won both houses of parliament (by a smaller margin) and the presidency with 61% of the vote. Iliescu's chief opponent, Emil Constantinescu of the Democratic Convention Alliance, received 39.5%. The elections in the fall of 1992 kept in power a coalition of former government supporters who were considerably less eager for economic and environmental reforms than their opponents.

Political reform received support from Phare after Romania's EU Agreement became operational on February 1, 1995. A legislative council was established to monitor the adaptation of its political system to European Union law.

The Civil Society

A revised political party system was encouraged by a decree of 31 December 1989 which set the threshold number of votes for participation in the parliament for both parties and coalitions. In 1992 only eight parties and coalitions out of 150 made it into parliament. Five separate configurations can be identified: the leading party, Social Democracy, which has allied with the Social Democratic party and the Democratic Party/Front of National Salvation, primarily former Communists; a revived liberal party; a coalition of Christian Democrats with the National Peasants Party as its backbone, a coalition the liberal party members left; a Romanian National Unity party in Transylvania, pro-reform but ultra-nationalist, and a party formed by Hungarians who constitute 7.1% of the population.

In the elections of 1996, the winning Democratic Convention included 15 groups: civic lobbies, greens, liberals, Christian Democrats, among others. The largest, the Social Democratic Union, supports Constantinescu's reform policies. It is believed that members of ethnic Hungarian groups in Transylvania also supported him (*The Economist,* November 23, 1996, p. 57).

Media

Freedom of opinion is guaranteed and censorship prohibited by the constitution. The Chamber of Deputies adopted a resolution supporting ethics policies applied to journalists by the Council of Europe. Almost all (92%) Romanians have television sets, with 28% able to use satellite or cable networks. The government controls the only nationwide television channel and has taken steps to prevent competition.

In the first days of the overthrow of the government in 1990, opposition newspapers were widely available. The main opposition paper, the Romania *Libera*, sold 1.5 million copies then but is now down to 140,000 because the government controls newsprint and has limited access to it.. According to a dispatch in the *New York Times,* there is only one newsprint plant in Romania "in an antiquated site 120 miles northeast of Bucharest," a handicap to all but the government owned press and a challenge to the new parliament and president (Perlez, The New York Times, June 8, 1995). This is a situation which the supporters of Constantanescu hoped would change as he assumed office.

Interest Group Representation

There are five large trade union federations which include a significant number of smaller unions. The federations represent labor interests at the national level in negotiations with the government. Trade unions are estimated to represent 60% of employees in industrial and commercial enterprises as well as farmers. There are trilateral dialogues between government representatives, employers and the unions regarding wages and other matters. Other organizations represent ethnic and religious groups, former political prisoners, human rights activists, environmentalists, and other segments of the population which have lacked political influence in the past .

Table 15.1 Economic, Social Indicators for the Danube River Basin Transition Countries, 1990 - 1995, Romania

INDICATOR	1990	1991	1992	1993	1994	1995
Social Indicators						
Population millions	23.2	23.2	22.8	22.8	22.7	22.7
Population growth rate %	0.5	0.5	0.0	0.0	0.0	0.0
Infant mortality rate per 1000 live births	19.0	18.0	22.0	21.2	19.9	18.7
Life expectancy (M)	69.0	69.0	68.0	68.3	68.8	69.3
(F)	75.0	75.0	74.0	74.3	74.8	75.4
Literacy rate %	NA	NA	97.0	97.0	97.0	97.0
Overall Economic Conditions	1990	1991	1992	1993	1994	1995
GDP in $ bn	38.2	28.9	19.6	26.4	28.1	33.7
Real GDP growth %	-5.6	-12.9	-8.8	1.5	3.9	6.9
GDP per capita in $	1,648	1,244	2,700	2,700	2,790	NA
Industry % of GDP	48.2	43.6	44.3	41.0	NA	NA
Agriculture % of GDP	18.0	18.5	20.1	21.0	NA	NA
Unemployment rate (in % of labor force, end-yr)	NA	3.0	8.4	10.2	10.9	8.9
Consumer price inflation %	5.1	174.5	210.9	290.3	136.8	32.3
Comm'l exchange rate (av) Lei:$	22.4	76.4	308.0	765.1	1,655.1	2,044
General government balance in % of GDP	1.2	0.6	-4.6	-0.1	-3.0	-2.0
Trade balance in $ bn	-1.8	-1.3	-1.4	-1.1	-0.4	-1.2
Current account $ bn	-3.3	-1.0	-1.5	-1.2	-0.4	-1.4
External debt in $ bn	0.4	2.0	3.0	4.0	4.4	5.3

15.2. Economic Policies

As a way of freeing the country from dependence on Moscow, Ceausescu in the 1980s initiated a policy of self-sufficiency, paying off international debts in March 1989, and following growth policies which used energy sources (oil, gas, hydro-power and lignite) at a rate three to five times that of western Europe. As the environment suffered, the distribution of food was irregular so that there were long lines for basic items like milk and bread. Building projects were begun in Bucharest and in other major cities and never completed. To increase the population birth control was not available. Poverty prevented families from looking after unwanted children and large numbers were placed in orphanages.

The Romanian economy, in Jackson's view, was uniquely distorted as a result of the Ceausescu rule. "The Ceausescu government imposed on the country a severe regime of import substitution for investment goods and consumer goods. In 1988 and 1989 domestic sources supplied virtually all standard investment goods...many investment goods were produced in tiny quantities with consequent high retail costs and low quality." As Romanian industry became increasingly self-equipped equipment breakdown was common and repairs difficult. The result: the Romanian economy in

1990 and since has been faced with one of the most severe restructuring challenges in the region, especially in heavy industry (Jackson, 1994, p. 555).

15.2.1. WESTERN TIES

Romania reached an association agreement with the EU on January 2, 1993; it has been in force from January 2, 1995. Romania's westward tilt in the early 1980s had earned it most favored nation (MFN) status with the US. A trade and economic cooperation agreement with the EC also provided reciprocal MFN treatment. The EU Association Agreement is similar to the agreement signed with Hungary but contains a human rights clause, as do agreements with the Czech Republic, Slovakia and Bulgaria. It provides for the consolidation of the earlier trade concessions and the establishment of a free trade area over a 10-year period.

A number of projects are under way under the Phare program to improve agriculture, transport, health and financial sectors, enterprise restructuring, privatization and human resources. Aid totaled 230 MECU by the end of 1992, 130 MECU in 1993. Romania has received humanitarian and emergency assistance as well as loans to overcome balance of payments difficulties (EC, March 5, 1995).

15.2.2. INSTITUTIONAL PROGRESS

Overall, the legal framework for restructuring the economy, while not complete, has been substantially achieved. The liberalization of prices began in late 1990; prices of over 90% of production and consumption goods are no longer under state control. The national currency was devaluated and made internally convertible in November 1992 while large industrial enterprises began to be privatized in 1992. Agricultural trade and service sector privatizatioon has lagged. Legislation passed in 1995 was to accelerate the introduction of a market economy, improve efficiency, privatization and to introduce bankruptcy laws. A stock exchange was developed in 1995-96.

The slow pace of state-directed restructuring was strongly criticized by Romanian opposition parties. The IMF delayed disbursement of a $265 million loan agreed to in May 1994; the World Bank provided a loan of $260 million in late 1995, conditioned on the restructuring of debt-ridden state companies.

Beginning in 1990 consumer and production prices were gradually liberalized and all price restrictions lifted except for those on bread and milk. A western-style tax system was established, and anti-inflation measures taken. A 1993 IMF agreement stabilized the monetary sector; inflation decreased from an annual 290.3% in 1993 to 136.8 in 1994, and 40.3% in 1995. According to EU criteria, Romania went from risk group "D" to "C" with respect to external debt, public debt and budgetary deficits.

Foreigners have the same personal and property rights as Romanian citizens, except that they cannot buy landed property. The private sector did not exist before December 1989 but now employs 47% of the working population, and constitutes 29.4% of exports and 32% of imports. Foreign investment is increasing.

15.2.3. MEASURING CHANGE

Marvin Jackson, Director of the Leuven Institute for Central and East European Studies, Catholic University of Leuven, Belgium, wrote critically in 1994.

> 'Transition' in Romania amounts to an impressive list of legal measures with less evident substance in either domestic market development or private enterprise. Nevertheless, numerous new bureaucracies have been set up, most of which provide their bureaucrats with both opportunities to influence new wealth creation and international connections. The system also serves the interests of a corps of western consultants, supported by international agencies...The Romanian people and Western supporters have so far paid little attention and received few results (Jackson, 1994, p. 553).

Over-optimism
The belief that Romania would fulfill all of the EU criteria by 2000 seems optimistic to many international experts. Before 1996 there was no clear break from the communist past. While many reform laws have been passed and programs put forward, at the informal level there was much behind-the-scenes manipulation; and privatization existed only at the retail level. During the Iliescu period, "Old networks still [held] the structures of command and interference in place, and large-scale privatization ha[d] barely started (Jackson, 1994, 576)." A slow transition has meant large decreases in output and high inflation, even higher than in neighboring Bulgaria.

Currency Problems
The Romanian central bank has limited yearly re-authorizations to trade foreign currency with four Romanian banks. Licenses of three large foreign banks to trade in currency were suspended in March 1996 (Perlez, "Romania bans currency deals by foreign banks," (*New York Times*, March 23, 1996). Foreign business executives in Bucharest say the move tends to decreases foreign investments, already slow compared with other EE countries. In contrast with Hungary and the Czech Republic the banks were politically controlled; the goal was to support the deteriorating leu. The Chairman of the Chamber of Deputies, Adrian Nastase, a leader of the Social Democracy Party, said foreign banks should not dictate the value of Romanian currency.

Collective farms were liquidated in 1990 and recently state farms have been transformed into stock companies. The private sector now controls about 70% of farm land and 80% of the land under cultivation. Farm output in 1993 was less than in 1989 partly due to drought conditions and the failure to increase irrigation sources.

15.2.4. SOCIAL CONDITIONS

Unemployment remains high in Romania and wages low. Official government figures put average annual income at $100 a month in 1996. Orphanages have filled again because of high contraceptive and abortion costs. The *New York Times* reported on March 25, 1996 that after reaching a low of about 80,000 institutionalized children in 1993, the number increased to 104,000 in 1995.

15.3. Environmental Issues and Policies

Like its neighbor, Hungary, for all practical purposes all of Romania is in the Danube catchment area. A significantly larger country, it occupies 29.3 % of the basin. Ceasescu policies of rapid, intensive development without regard to environmental protection caused severe damage to natural resources. Most rivers remain severely polluted and lifeless over long distances. Toxic emissions of oxides of sulfur and particulates from energy producing sources are well over acceptable standards. Sanitary disposal of solid waste is minimal and the management of hazardous waste far below standards. Ground water is polluted in several areas and non-point source runoff from agricultural chemicals contributes to river pollution. The coastal area of the Black Sea, a thriving tourist area in the past, discharges untreated municipal and other waste into the Sea.

15.3.1. STATUS OF ENVIRONMENT

Water Resources

In a large number of water supply districts elevated nitrate levels have been found; in 14 districts with higher levels up to 13% of the newborns each year developed methemoglobinemia (Hertzman, 1995, p. 42). Although DDT had been banned in Romania it was still being produced for export in the 1990s. Carcinogenic substances including chlorinated pesticides have been found in tap water samples taken in the southeastern part of the country and along the Prut River.

Romania's poorly designed and improperly used irrigation methods have had significantly unfavorable impacts on the environment. Irrigation facilities could provide water for 3.2 million hectares if they worked properly. Instead their deterioration limits irrigation to 2 million hectares. Pumping costs are high. Energy use for irrigation constitutes about 5% of the country's power consumption. The cost must be fully subsidized by the government.

Water Supply

Sixty-six per cent of the total water supply comes from within Romania, 22% from Bulgaria, 7% from the former Yugoslavia (Serbia), and the remainder from Ukraine. The Danube forms the boundary between Romania with Bulgaria (471 km), the Ukraine (134 km) and the former Yugoslavia (94 km). There are about 550 impounding reservoirs and more than 130 dams higher than 15 m in the country.

There has been a decrease in the total amount of water consumption since 1991 as the result of the slowdown in irrigation and industry associated with the economic transition. There are two large hydropower plants on the Danube, Iron Gates I and II. The annual hydropower potential of the country is estimated to be 38 billion kilowatt, of which the Danube can theoretically provide 26%. In 1990 hydropower contributed 17% to national electric production. Only 15% of the rural population is served by water networks. The periodic failures of the tap water supply remain a serious problem for the third of the population which uses it. Severe problems continue with the quality of Romania's drinking water. High levels of pesticides have been found in tap water samples from over 400 towns, with the highest levels found along the Danube. Elevated nitrate concentrations were found in local water supplies in 39 of 41 districts (information from IUCN, 1994, p. 166).

Biodiversity

Romanian forests cover an area of 6.6 million hectares, almost entirely in the Carpathian mountains or their foothills. Elsewhere intensive agricultural use has cleared most of the former forested areas. Dryness caused by climatic conditions as well as pollution have also had a serious impact on forests. There are three biosphere preserves, of which the Danube Delta, comprising over 590,000 hectares, is by far the largest. There are 12 national parks and a number of protected areas. Management plans and supervision are lacking, especially in the protected areas.

Protection of the Danube Delta ecosystems is important for two primary reasons (IUCN, 1992). They have traditionally served as an environmental buffer between the Danube and the Black Sea, filtering out pollutants and sustaining water quality and the natural habitats of fish in the delta and the environmentally vulnerable shallow waters of the northwestern Black Sea. And the delta is Europe's largest remaining natural wetland, with unique ecosystems, home to many rare bird species. There has been a sharp decline in its value as a biological buffer and wetland ecosystem over the last 40 years. Increased pollution loads from the Danube, especially phosphorus and nitrogen, have resulted in eutrophication of some of the areas 668 lakes and the accumulation of toxics and heavy metals in soils and fauna. Construction of upstream dams and the filling and destruction of Danube floodplains and wetlands have resulted in interference with the delta's natural hydrological cycle.

All of these phenomena, plus poorly planned and constructed canals, dikes and polders and ineffective management of economic activities like tourism, hunting and farming, have interfered with the water regime and destroyed wildlife habitats and fish breeding grounds. Reed harvesting and forestry activities have also been poorly managed. The population has declined steadily as sources of livelihood have decreased; and drinking water problems multiplied.

The Strategic Action Plan has recommended two projects, one for the entire delta to provide management improvement as well as the implementation of a detailed management plan. The lower Danube floodplain between Calarasi and Galati, a protected area of 15,000 ha. will be studied for improved management and restoration.

Other Resources

Because of a lack of data on sediment transport regimes in all parts of the Danube, it is difficult to identify changes in them and the damage that has been caused in the riverine ecosystem. The Iron Gate dams have created difficult problems. About ten per cent of the 70.7 million tons of suspended material transported each year to the delta are retained there while the remainder is discharged into the Black Sea. Data and information on impacts, causes and possible solutions are not close at hand.

Romania's national air monitoring network records higher than admissible levels of toxic waste primarily from the burning of soft coal for heating and extensive non-ferrous metallurgical plants which generate lead, zinc and other heavy metal pollutants. Chemical and petrochemical industrial centers cause contamination as do pulp and paper mills and artificial fiber factories; chemical fertilizer plants, and thermal power plants using fossil fuel in many cities including Bucharest and Constanta.

Pollution Sources

Sewer networks are inadequate. Sludge treatment involves drying beds or lagoons in smaller communities and anaerobic digestion in larger localities. Planning and implementation of treatment facilities have been poor. The largest population centers, including the capital Bucharest, and the cities of Cluj and Brasov, discharge their waste untreated while sizable funds have been spent to improve waste treatment in smaller settlements. Lack of funding has put a halt to a plant in Bucharest which is 70% completed. An estimated $100 million would be needed to complete it.

Over half of the total land area is classified as agricultural. About 67% of this, or 10 million hectares, are cultivated and 4.7 million hectares are in meadows and pastures. Climate and fertile soils supported a flourishing agriculture until the 1980s. Until then domestic food needs were met and Romania was a large net exporter. A number of unfavorable factors caused a sharp decline in output in the 1980s, especially the deterioration of the irrigation system.

Large scale cultivation on cooperatives and state farms is the rule although there has been some privatization under way in accordance with the Land Reform Law enacted in 1991. Soil erosion, waterlogging and salinity, mining and dumping are all threats. Erosion can be corrected with proper soil conservation measures and this has occurred in some areas. Pollutant discharges from industry affect about 300,000 hectares with pollution from heavy metals predominant.

Increasing nitrate concentrations in groundwater and organic and inorganic micro-pollutants, especially pesticides and heavy metals from agricultural runoff and livestock wastes are known to be contaminating the water supply. Monitoring is limited; a large data gap makes assessment and planned treatment difficult. Romania lacks suitable regulation of land waste disposal. More than 1,900 areas are covered with deposits from communal animal waste, farms, mining and oil residue, chemical, metallurgical, and construction materials. Standards or regulations for use, movement and disposal of hazardous waste and design criteria for disposal facilities, including medical waste, are lacking. Most land fills are operated in an unsatisfactory manner. Hazardous waste has recently been imported into Romania without proper transport, storage or disposal control. Industrial waste disposal on dumps, ponds, lagoons also needs regulation.

Engineering Structures

The hydroelectric power projects at Iron Gates I and II and the Danube-Black Sea canal are the two most prominent water projects in the Romanian reach of the Danube. The canal was opened in 1984 to permit navigation through the Delta to the Black Sea. The reservoir created at the Iron Gate I hydropower plant (2,100 MW) is about 117 km. in length and contains about 3.5 billion m³ of water. The Iron Gate II hydropower plant was built in 1980 with a capacity.of 432 MW. The long-term effect of large water project structures is not well understood, although studies have shown that fish migration and sediment transport are seriously affected with unfavorable impacts on habitats along downstream river banks.

Health and Hot Spots

Water supply failures combined with the presence of faecal contaminants have created severe health problems throughout Romania. Pre-investment studies have disclosed a number of problems. While neglect of environmental protection policies during the Communist period affected natural resources in all parts of the country, a number of "hotspot" areas have been identified. Several of them have been caused serious public health problems. Naia Mare, Copsa Mica, Medias and Zlatna, have experienced respiratory diseases from pollution higher than those in areas without air pollution. Children in these areas were twice as apt to suffer from chronic upper respiratory diseases and malnutrition. Data on occupational health problems in mines and within plants show high levels of respiratory disease. The Baita uranium mine in Bihor has lung cancer rates 2.5 times control communities.

Data on health conditions are the result of occasional, random studies. Structured studies are needed, in keeping with a set of national health goals. Some major tests have begun but this is an area that needs careful attention for short and long term improvement in public health.

The Strategic Action Plan has identified forty three clean-up projects for remediation in especially polluted areas; 30 of them are considered to be high priority. Fifteen wastewater treatment plants are in need of replacement or modernization, 3 of them high priority. A number of plant modernizations which will lower water pollution are also listed as high priority including pulp and paper mills, petrochemical plants, and fertilizer factories, among others (EPDRB, 1995a, p. 46-47). In contrast with the Hungarian projects, most of which had been studied and costed for possible implementation, none of the proposed Romanian projects include estimated costs.

15.3.2. GOVERNMENT POLICIES

The new constitution of November 1991 established a more supportive framework for environmental legislation, including guaranteed rights of access to information of public interest and the affirmation of the obligation of public authorities to provide correct information to citizens. It also stipulates that the state must secure environmental protection and recovery; as well as preservation of ecological balance. The country is divided into 41 counties for administrative purposes.

Framework legislation enacted in 1973, the Environmental Act, remains in force pending the approval of a new principal environmental law which has been passed by parliament and remains to be promulgated by the president. An environmental strategy and National Environmental Action Program was adopted in December 1995. New ordinances have been introduced during the transition. The proposed new Environmental Act is wide ranging and includes specific laws dealing with the creation and management of the Danube Delta Biosphere Reserve, waste water treatment and water. Delays in the promotion and adoption of the environmental law have created pressing problems.

A number of EC council directives have as yet not received legislative support. While their objectives are included in the environmental plan of action adopted in 1995, incorporation into national legislation has not been achieved. This includes the polluter pays principle, environmental impact assessment, effective access to public information among others. With respect to water quality, fundamental legislation embodying the EU directives has not yet been passed. There are no standards for bathing water, for example, and no emission standards for waste discharged into surface water. Nor are there any regulations for the protection of ground water (Habcu, 1996, p. 103).

Key Institutions

Although other ministries (Health, Public Works and Land Planning, Labor and Social Protection, Agriculture and Food, Justice) have responsibilities for environmental management, the main environmental responsibilities reside with the Ministry of Water, Forestry and Environmental Protection (referred to as the Ministry of Environment) which replaced the Ministry of Environment established in 1990.

The Ministry of Environment is responsible for the establishment of all media standards and their enforcement. There are problems with the standards themselves, the way they have been monitored, and the way they have been enforced. The Ministry of Environment and the Ministry of Health have consistent but overlapping monitoring programs. Collection of samples and their analyses are limited by outmoded technology and poor system control.

The Danube Delta Biosphere Reserve Authority is a specially created body responsible for environmental administration in the Delta region. The Ministry of Environment also works with nongovernmental organizations and other groups, such as the National Commission of Standards, Meteorology and Quality. The Research Institute for Environmental Engineering (ICIM) is responsible for the secretarial activities of the Bucharest Declaration, among other research activities. Local public administrations and environmental protection agencies play a role in each of the counties of Romania. These include the Public Local Administrations, the Environment Protection Agencies, the Romanian Waters Authorities, and the Mayoralties in each county.

15.3.3. POLICY IN TRANSITION

An environmental strategy paper developed by Romanian and World Bank experts in 1992 has been proposed as an action program for Romania. The National Environmental Action Plan will be based on the principles of the EU supported policies such as the protection of human health, sustainable development, the prevention of pollution, conservation of biodiversity and cultural and historical heritage as well as the polluter pays principle. The plan will guide legislation and its implementation. Much legislation remains under consideration within the Ministry. Western funding has been made available to help draft legislation and for other minimal activities related to environmental policy development. Much more may be available to the Romanian government from the Phare program, primarily through private western consultants, for the elaboration of policies recommended by the funders.

CHAPTER 16

Slovak Republic

16.1. Slovakia under Meciar

The post-1992 development of the new republic of Slovakia has been different in many important political and economic respects from that of the Czech Republic. The twice-elected prime minister, Vladimir Meciar, has been accused by critics in the west and in eastern Europe of violating democratic processes in order to retain political power. Legislation proposed by Meciar would centralize power in his hands, restrict the functions of the constitutional court, and violate the principles of free speech. The EC has officially complained about human rights abuses; the European Parliament adopted a resolution criticizing the Slovak government for its anti-democratic actions. It has issued demarches noting that Slovakia may not make the first round in acceptances into the EU from Eastern Europe.

Economic development has suffered because of the slow approach to privatization adopted by Meciar and the reluctance of foreign investors to provide support for industrial and commercial enterprises (Kinzer, *New York Times*, Dec. 26, 1995). The three ruling parties are accused of basing official appointments at the local level on political party affiliation rather than ability.

16.1.1. WESTERN ORIENTATION

Slovakia continued membership in the major international organizations after its separation from the CR. An EU Agreement entered into force in February 1995. While EU, WEU, and NATO participation is identified by government leaders as basic to Slovak foreign policy, many decisions and public pronouncements reflect serious misgivings about complete integration with the west. Promised domestic economic improvements have not yet been realized. Added to this is a latent nationalist feeling on the part of a significant portion of the population which has been fanned by Meciar and his followers.

Slovak foreign policy, in the words of one commentator, "is based on an exaggerated opinion of its national capabilities as well as on a special ambivalence rooted in wishful thinking (Bricke, 1996, p. 205)." At the same time nationalist phobias jeopardize the development of a common bond with the CR and cause hesitation about integration into the European market. The fact that 95% of Slovak energy imports (oil and gas) continue to come from Russia has tended to lessen the importance of economic integration with the west.

The Phare help which followed the signing of the Europe Agreement included funding assistance for small enterprises and for related training and research, as well as the reconstruction of its energy, transport, and telecommunications infrastructure. The leading-party newspaper has called such help large-scale economic espionage. Ensuing dialogues between EU representatives and Slovak leaders have been going slowly: "[The] intensity of contacts leaves a lot to be desired inasmuch as the Slovakian political class is quite unfamiliar with democratic norms and political declarations often are only lip-service (Bricke, 1996, p. 220)." Slovak failures to meet OECD criteria will probably delay its consideration until after 1997.

Controversies with Hungary

Controversies with its neighbor Hungary have included the refusal of the Hungarian government to continue to support the downstream part of the Gabcikovo Dam System. The part completed by Slovakia diverts the water of the Danube River via a 17.5 km. bypass canal to an 8-turbine hydroelectric power dam. A dispute over the legality of the respective actions by both countries is now before the International Court at the Hague (See Fitzmaurice, 1996).

Slovakia and Hungary signed a basic agreement concerning the status of the Hungarian minority in Slovakia in the spring of 1995. It was ratified by the Hungarian parliament in June 1995. Slovakia failed to ratify the treaty until April 1996 and then only after the parliament had adopted measures considered to restrict the rights of the Hungarian minority.

16.1.2. CONSTITUTIONAL AND INSTITUTIONAL CHANGES

The constitution of the new republic was adopted by the National Council in September 1992. The legal system thus established is essentially the same as that in place immediately before the separation from the Czech Republic. Critics have commented, however, that the new constitution seems to provide for a revival of the spirit of national paternalism. Its definition of national property is so broad that almost all property can be identified as "national," which in turn can block the privatization process and restrict the activities of private entrepreneurs. While parliament can dismiss the president for political reasons, it is also possible for the president to dissolve the parliament. This feature makes it difficult for the president to mediate a serious political dispute. Its provisions do not meet the provisions of the treaty of Maastricht. There are fears that Meciar wants to turn the parliamentary system into a more centralized presidential system.

The Slovak Parliament approved a law on March 27 1996 which would allow the prosecution of individuals who "spread false information" that could damage Slovakia's interests or who organize public rallies "with the intention of subverting the country's constitutional system, territorial integrity, or defense capability." Critics of the measure include the Roman Catholic Church, trade unions, judges, and civic organization all of whom agree that the vaguely worded legislation could endanger freedom of expression and assembly (OMRI 27 March 1996, p. 1). Slovak President Michal Kovac vetoed the legislation on April 4, 1996. In July the Parliament approved legislation to divide the country into eight regions and 79 districts. The move was designed to enhance Meciar's political powers by reducing the independence of some cities, including the capital of Bratislava and the source of opposition to him. It was criticized as costly and detrimental to the administration of environmental protection policies (*OMRI Daily Digest*, July 8, 1996).

The Civil Society, Political Parties

Since the third set of free elections in Slovakia, held in September 1994, the party system has been fragmented and polarized. Meciar has the support of a coalition which won 82 out of 150 seats in 1994. His party, the Movement for a Democratic Slovakia (HZDS), and its allies, the Slovak Workers Association (ZRS) and the Slovakian National Party (SNS) gained 48% of the vote. The leaders of all three parties have attracted followers strongly oriented to national, social and egalitarian values and represent them in a confrontational style. The "programmatic" parties, Christian Democrat, the Party of the Democratic Left, and the Hungarian coalition, won 38% of the vote. Polarization and constant political infighting between the ruling coalition and its opposition have had a destabilizing effect on governance.

Media

The coalition government has a monopoly over state television and broadcasting contrary to existing law which provides that they should be publicly run. All eighteen members of the boards of the Slovak Radio and Slovak Television were replaced in November 1994, following the election of the Meciar coalition (Butora, 1995). Meciar supporters in parliament were named to head the state radio and television systems. News reports on public television are accused of bias and often seem to be similar to those published under socialism. Leading newspapers constantly fight against government attempts to curtail free speech.

Interest Group Representation

The number of NGOs has increased significantly since 1989. Where there were once only a few government-subsidized groups there are now over 10,000 nongovernmental organizations representing health, public welfare, education, youth, environmental and minority interests. Western NGOs have been an important stimulus in this development and provided help to these nascent groups. The Meciar government has attempted to control several of the larger organizations.

Legislation passed by parliament on June 20, 1996 curtailed the rights of private foundations by reequiring basic assets of 100,000 crowns ($3,200), prohibitively high for small foundations, which are also required to register with the Interior Ministry rather than with independent courts. Law supporters said foundations were politically active and actively opposed to the government. The Soros Open Society Fund and its founder, George Soros were particular targets (Fisher, 1996).

16.2. Economic Policies

16.2.1. IMPACT OF SEPARATION

The SR economy was severely affected during 1991 and 1992 as the CMEA market for its military and other goods disappeared. The CSFR had been the largest supplier of weaponry in Eastern Europe for the Warsaw Pact countries. Disparities between the two republics in unemployment and other indicators created tensions in the early part of the transition period. The federal assembly could not reach agreement on economic reforms. The immediate disadvantages for Slovakia were more severe than had been contemplated. By and large advantages were on the CR side following the separation since it reduced the number of military enterprises and decreased its dependence on large industry.

Drop in Trade, Investment

Independence has meant not only a sharp drop in trade between Slovakia and the Czech Republic but contributed to high unemployment and low foreign investment in Slovakia. In splitting from Slovakia, the Czech Republic left behind the federal republic's least successful industries: the gigantic steel and textile mills and weapons production plants. Slovakia had an appreciable increase in industrial development during the late 1940s, with an output almost on a level with the CR. "The sectoral emphasis reflected the regime's obsession with heavy industry, exemplified by the increase in the index of output of ferrous metallurgy from a base of 100 in 1948 to 4,144 in 1989...(Brada, 1994, p. 519)."

The share of national investments in agriculture decreased sharply, from 30% to 7%, in the same period. Slovakia's increased chemical sector was based largely on oil imports from the USSR. It was also heavily integrated into the Soviet defense sector through its supply of weaponry and defense equipment (Brada, 1994, p. 520). Independence initially resulted in a sharp drop in trade with the Czech Republic and a severe economic recession with high unemployment and low foreign investment.

Table16.1 Economic, Social Indicators for the Danube River Basin Transition Countries 1990 - 1995, Slovak Republic

INDICATOR		1990	1991	1992	1993	1994	1995
Social Indicators							
Population millions		5.3	5.3	5.3	5.3	5.3	5.4
Population growth rate %		0.3	0.3	0.2	0.5	0.5	0.5
Infant mortality rate per 1000 live births		12.0	13.2	12.6	15.6	11.2	10.0
Life expectancy	(M)	69.0	69.0	68.0	68.2	68.7	69.2
	(F)	76.0	77.0	76.0	76.9	77.2	77.6
Literacy rate %		99	99	99	99	99	99
Economic Indicators							
Overall Economic Conditions		1990	1991	1992	1993	1994	1995
GDP in $ bn		NA	NA	10.9	12.1	14.2	17.4
Real GDP growth %		-2.5	-14.5	-1.6	-4.1	4.8	6.6
GDP per capita in $		NA	6,900	6,100	5,800	6,070	NA
Industry % of GDP		61.6	63.9	38	36.7	37.5	NA
Agricultural % of GDP		8.2	5.8	6.1	6.6	6.1	NA
Unemployment rate (in % of labor force, end-yr)		NA	6.6	11.4	12.7	14.8	NA
Consumer price inflation %		10.4	57.8	10	23.2	13.4	9.9
Exchange rate (av) Kcs/Sk:$		18	29.5	28.3	30.8	32	29.8
General government balance in % of GDP		0.1	-2	-13.1	-6.7	-3.7	-3
Trade balance in $ bn		-0.8	-0.4	-0.7	-0.9	0.1	n/a
Current account $ m		-949	329	316	-679	650	451
External debt in $ bn		7.6	9.1	1.9	3.2	4.2	NA

16.2.2. INSTITUTIONAL PROGRESS

The radical reforms which took place from 1990 to 1993 followed the practices of other former communist bloc countries: far-reaching price liberalization, three-step currency devaluation, the liberalization of trade regulation and introduction of a tight money policy. The government budget was stabilized by cutting state subsidies for utilities and other services and increasing rates for their use.

The first moves toward privatization using the voucher method were initiated in the CSFR. Under this system all citizens could buy a set of investment points for the equivalent of $33 which would constitute shares in selected former state enterprises. Upon Meciar's return to power in 1994 the second wave of voucher investments was replaced with an issue of government bonds. The Meciar government has been accused of selling state enterprises at low cost to those especially loyal to his party, thus not only benefiting supporters but also reducing income to the state. Some utilities and armaments manufacturing enterprises have been completely or partly exempt from privatization. The banking sector is still largely state-controlled with several weaknesses which have yet to be effectively addressed. Pre-communist period loans of a questionable nature are 20% of banking assets.

16.2.3. MEASURING CHANGE

The wisdom of the separation from the CR instigated by the twice-elected Premier Meciar continues to be debated, especially in urban areas of Slovakia, where the loss of trade added to a general post-1989 recession. Disadvantages for Slovakia included its inheritance of the federal republic's least successful industrial complexes which had been built in the 1960s to accommodate Soviet central planners. Slovakia has had problems developing a consistent privatization program, unlike its Czech neighbor. Its largest companies were geared to highly specialized markets for heavy engineering products in CEE and the Soviet Union. Between 1988 and 1992 military production, about 32% of Slovakian engineering industry declined by 90%; most of 40 major plants shut down with a loss of 60,000 jobs. In 1994 the OECD reported that both countries suffered from the split with adjustment costs, however, much higher for Slovakia.

Positive Signs

There were positive signs for the economy in Slovakia by the end of 1994. More than half of the GDP was created by the private sector and experienced modest growth. Exports to OECD countries rose, with half of total exports going to them. Its rate of growth placed Slovakia fifth in a survey of 25 post-Communist countries by the *Wall Street Journal* (Butora, 1995). Standard & Poor's credit rating for Slovakia was increased so that it equalled that of Poland and was ahead of Hungary.

Macro-economic developments were praised as the economy grew by 7.4% and annual inflation was only 7.2%. Standard & Poor warned that further upgrading may be impeded by the country's unstable political situation and lack of industrial restructuring. The agricultural share of the GDP fell by 50% to 6.5% in the first two years of the transition. Dwindling production, partly the result of shrinking demand and restrictive monetary policies, but primarily of very large subsidy cuts and farm land fragmentation. Additional constraints include general economic uncertainty and low investments in the agricultural sector.

16.2. 4. SOCIAL CONDITIONS

Devalued pensions, unemployment, and inadequate health and other social services continue to take their toll on living standards for most of the Slovak population. Crime, exacerbated by a corrupt police force, is steadily increasing under the guidance of a mafia type of organization. Racism on the part of groups of "skinheads" who attack Roma is an increasing problem.

16.3. Environmental Issues and Policies

Inadequate treatment of domestic and industrial waste caused significant increases of pollution during the Cold War. The effects have been summed up in a report on the status of its National Environmental Action Program:

16.3.1. BASIC ENVIRONMENTAL ISSUES

Wasteful exploitation of natural resources, extensive pollution of the air, water and land, release of contaminants into the environment and the food-chain, ill-advised intrusions on the countryside, the accumulation of waste, and the distorted structure of the Slovak economy (with its outdated technology and infrastructure) have combined to bring about a comprehensive deterioration in the state of the country's environment, with various degrees of devastation and risk in different regions (Klinda and Fischerova, 1995, p. 95).

Slovakia is marked by valleys and canyons which sometimes provide insufficient natural ventilation for air pollutants. In 1991 annual emissions of sulphur dioxide were among the highest in Europe, putting Slovakia in fifth place among European countries (behind only East Germany among the CEE); Slovakia was in tenth place for emissions of nitrous oxides. With respect to water resources, the priority problems identified in the National Action Plan (prepared for the Danube Program as a follow-up to the Strategic Action Plan) were eutrophication of surface waters, pollution of surface waters from both point and non-point sources, a failure to manage solid waste, and a decrease of ground-water levels.

Gabčikovo Project

The Gabčikovo hydroelectric power project put into operation by the Slovak government has created world-wide attention. The dam was built despite the fact that experts severely criticized its potentially detrimental impact on water supply and wetlands and that it was difficult to justify its cost to the environment and to the country on the basis of the presumed benefits it might bring in energy or other improvements such as improved flood control.

Natural Resources

Almost all of Slovakia is located in the Danube Basin. Sixty per cent of the country's population of 5.3 million is concentrated in western and northern Slovakia in the Lower Morava and Danube basins and eight sub-basins. The most heavily populated are the Vah which is also the largest in area (1.4 million), the Nitra (.7 million) and the Hornad (.7 million). Fifty per cent of the land area is agricultural with 30% arable; 40% consists of forest ecosystems. the rest is developed urban or rural settlements. Of the agricultural area, arable land is evenly distributed in the Danube and lower Morava (more than 80%), Nitra (82%) and Vah (71 %) sub-basins (IUCN, 1994, p. 177).

Water Resources

Slovakia is able to meet its water needs through a well developed network of surface watercourses. There is also an extensive network of reservoirs. The existing network of surface waters permits water-transfer between sub-basins and is considered to be a well balanced network which meets current water needs.

Monitoring has identified eutrophication as the most serious water pollution problem resulting primarily from a high nutrient load from agro-chemicals and the insufficient capacity of sewage treatment facilities. Ground water is polluted as a result; a high percentage of wells do not meet drinking water standards. Eutrophication also occurs in water reservoirs, primarily in municipal sewage. Phosphorous from detergents is not monitored and there is no fee charged for releasing it; sewage systems are not well equipped to eliminate these nutrients. Increased salination from agricultural and industrial sources, heavy metals from industrial and municipal discharges are also of concern.

Other Resources

Many ecosystems "are characterized by very rich and complex ecological structures which despite human impact (landscape changes, over-exploitation, pollution) still maintain levels of biodiversity among the richest in the Danube river basin (IUCN, 1994, p.180)." Natural ecosystems, affected by a decreasing and more polluted water supply, are deteriorating. A wetlands survey is under way; four Ramsar sites have been identified for protection in the Danube catchment area. The Strategic Action Plan recommends six projects for biodiversity restoration. Networks of protected areas were designated during the last century, two parks, Priboj and Ponicka dubrava having been established in 1895. Laws passed in the mid-1950s began a systematic set of designations of such areas.

Authorities have begun a national survey of river and lake sediments which has so far revealed above standard concentrations of heavy metals, PCBs and DDT among other contaminants. The former Czechoslovakia was one of the heaviest polluters of the atmosphere in Europe. Twelve regions in Slovakia with 24% of the population have significant air pollution. Sources include the burning of low quality coal, metallurgy plants, chemical plants, as well as the cellulose and paper industries. Cars without catalytic converters are also sources. The Clean Air Act of 1991, amended in 1993, established new emission standards. Other measures are being taken to correct air pollution problems.

Soil contamination assessments are based primarily on level of fertilizer use. There is a general lack of information concerning levels of pesticide use in the different agricultural areas. Due to rapid and extensive soil erosion, surface waters are loaded with high amounts of nutrients and particulate organic matter (IUCN, 1994, p. 182).

Pollution Sources

High levels of industrial pollution have been found in the Danube, Morava and other basins. Heavy metals and untreated organic and inorganic contaminants as well as waste from a refinery and petrochemical plant pollute the Danube. Twenty per cent of the total wastewater from industrial activities in the Morava Basin is treated by municipal plants; some of the remainder is treated within plants, the rest goes into the Morava untreated.

Similar situations exist elsewhere. The low percentage of contaminants (20%) treated by municipal treatment plants are not efficiently processed because of the low hydraulic capacity of the plants. Timea Spitkova, *Prague Post*, reported that the promise to treat all waste from the IstroChem Company, one of largest chemical plants in the country which is located in Bratislava, before it entered the Danube has not been kept. The cost of the technology that would be installed in the plant was estimated at $15 million. The report states that rejection by the government of privatization of the company foreclosed the possibility of funding for the proposed treatment technology. Privatization had been a condition of the EBRD loan to support modernization (Spitkova, 1995, p. 18).

Slovakia has a long history of intensive industrialization; industrial activity shows a close relationship to the network of watercourses and natural resources. Polluting industries include chemical works, engineering, food, pulp and paper, ferrous and non-ferrous metallurgy, cement, power-generation, textile industries, leather, timber processing and mining. While Slovakia is party to a number of international conventions related to environmental protection, drastic problems like air pollution caused by heavy industry, water pollution from lack of two-stage sewage plants, and toxic pollution in agriculture from pesticides, fodder, fertilizers have not been addressed "because financial means to fulfil commitments are neither available nor offered by the EU (Bricke, 1995 p.220)."

Municipal Waste

Municipal sewage outlets have a major impact on pollution. The larger municipalities have public sewage systems, and the wastewater discharged from these systems is mixed, with industry producing a significant proportion. Industrial complexes are responsible for up to 53% of total discharges the Vah sub-basin and up to 44% of total discharges in the lower Morava and Danube sub-basin. A significant proportion of existing sewage systems is directly linked to surface water courses; discharges enter the river network without treatment. Municipal discharges in the Danube sub-basin flow through 14 sewage systems into surface streams.

Many of the existing municipal sewage treatment plants are hydraulically overloaded thus high levels of wastewater may be discharged into rivers before treatment. The presence of specific contaminants, most of which are industrial in origin, further reduces the efficiency of treatment processes. Sludge produced within the sewage treatment plant concentrates high amounts of toxic heavy metals rendering it unsuitable for agricultural use. In most cases large amounts of sludge are deposited in unsecured landfills, close to cities and industrial complexes, and on river banks in floodplains.

Agriculture, Solid Waste

Agricultural development after WWII in Slovakia began with the establishment of state farms and agricultural cooperatives. The industrialization of agriculture was intensified from 1970 to 1989. Attempts were made to alleviate environment impacts but have been acknowledged by government reports to be ineffective. A number of potential problems with the use of synthetic pesticides and herbicides have been cited.

Prior to 1989 there were no attempts to manage solid wastes in the SR. Registration of disposal sites, required by law, has resulted in the identification of over 7,000 such sites. Only a small number fulfil legal requirements. Landfills are now under review and their content and potential harmful effects recorded. The Slovak Ministry of Environment reports that a number of landfills are sources of dangerous and special wastes and are registered and territorially localized. Landfills are concentrated especially in the regions of Bratislava, Galanta, Senica, and other major settlements near industrial areas. Operating incinerators have insufficient capacity to handle waste and do not meet air pollution requirements.

Structures

Dams have been built in several of the rivers in the Danube catchment area to control water balances, improve water resource availability, facilitate power generation and meet the increasing water demand from industry, irrigation and municipalities. The need for flood control was also a significant factor. Schemes include the construction of large and small dams, floodplain drainage, and water-transfer projects between sub-catchment areas. A long-term but still uncertain impact on the wetlands, biodiversity and biogeochemical cycles of the main elements may be expected (IUCN, 1994, p. 184).

Presently Slovakia has one working nuclear plant, Jaslovske Bohunice, which has four Soviet type reactors. These were put into service from 1978 to 1985. The safety of this plant is in question and it is not regarded as up to western European standards. Another nuclear power plant, the Mochove, which has raised controversy, and has been reviewed by western experts, is under construction.

Health and Hot Spots

Nitrates in drinking water in some rural areas cause significant health problems. Incidence of methemoglobinemia were cut back as the result of a public education program which included the supply of nitrate-free bottled water to mothers and infants. The Strategic Action Plan recommends work on 22 projects, with six identified as high priority. Of the 22, seven are for waste water treatment plants, three are high priority. A number of modernizations are planned including work on a tannery, thermal power, pulp and paper, and sugar plants. None of the projects has been costed.

16.3.2. GOVERNMENT POLICIES

As in most other east European Danube countries, the Slovak constitution adopted in 1992 granted the right to a clean environment to all while the state acknowledged responsibility for its protection. The constitution gave the state control over the use of natural resources, maintaining ecological balances and the general state of the environment. All had a right to information about the state of the environment. Framework legislation included a number of EC doctrines. General framework legislation, the Environmental Act, was passed in 1992.

Following the Lucerne Conference an environmental policy strategy was adopted with short, medium and long term objectives. It is being elaborated by district environmental offices. Harmonization of Slovak legislation with EU directives is also under way, including legislation providing for the use of environmental impact assessments before construction and certain other specified activities. A State Fund of the Environment was established in 1991 which is financed by fines and fees charged in connection with the use of various natural resources.

Updating of the Czechoslovak Federal Republic Act on Waters (1973) is anticipated. Regulations cover water withdrawals and treatment to maintain a balance between consumption and capacity. They provide for the protection of recreation, shipping and other important interests. Regulations have updated water quality standards (1993); new legislation is expected to incorporate these and other policy changes (Gasparikova, et al, 1996, p. 115).

Key Institutions

Three ministries have responsibilities for water supply and quality in the SR: the Ministries of the Environment, Soil Management, and Health. Even while joined with the Czech Republic, Slovakia had a separate government structure, with its own premier, cabinet, and national assembly. While there were some government structures in Prague which served the combined republics, including a federal ministry of environment, much of environmental policy implementation was conducted within each republic.

A 1990 Act of the Slovak National Council established the Slovak Commission of Environment which became the Ministry of Environment in 1992. The ministry is responsible for nature conservation; the protection of water quality and quantity and its use; air protection; territorial planning and building orders; waste management; monitoring, including the development of a uniform environmental information system; geological research and survey. The ministry also prepares regulations in the environment field and directs district inspectorates of environment. Monitoring of a range of environmental media is accomplished by the Slovak Hydrometeorological Institute

District environmental authorities have been given increased authority over various aspects of the regulatory function. There are 38 such districts. The Slovak Inspectorate for Environment, established in 1990, conducts water, air and waste management inspections. In June 1993 a Slovakian Agency for the Environment was established to implement and coordinate monitoring, information systems and administration of protected areas.

16.3.3. POLICY IN TRANSITION

The Slovak Ministry of Environment has defined its urgent tasks as amending environmental legislation, completing monitoring and information systems on the state of its environment, analyses and conclusions about the influence of the new technologies to manage the environment, proposals for forecasting environmental development and protection, including consideration of international cooperation, and realizing a number of pertinent environment improvement programs. Approximation of EU law is proceeding under the guidance of the Ministry of Environment.

The government pledged support for a number of budgeted activities such as handbooks to make understanding of environmental policies more widespread, imposition of ecological taxes on hazardous wastes, development of a method for assessment of ecological loading capacities of land, a monitoring system for reporting on conditions in the national forests, an ecological information system, and so forth. Legislative and administrative actions to carry out environmental goals is in a transition phase.

The current restructuring of the political system and the lack of funds are causing some delay in the implementation of national environmental action plans. Unfortunately the newly independent municipalities have serious difficulties with the establishment of relevant water works and sewerage. entities. Clean-up of serious existing water pollution problems is not economically feasible. The state water authorities are preparing a proposal which provides that the polluter be made responsible for rectifying damage which it caused, in return for certain tax concessions.

CHAPTER 17

Slovenia

17.1. Political Reform

When the Communist Party began to lose control of the Yugoslavian Federation in the 1980s Slovenia was a leader in the breakaway movement among the member republics. Beginning in 1988 a number of alternative political parties were established in defiance of a communist ban. Two Social Democratic parties, a peasant party, and a Christian Democratic Party were formed. In 1990 the Slovenian branch of the Yugoslav Communist Party unilaterally seceded from the party and announced its intention to hold multiparty elections. Free elections took place in April 1990, bringing to power DEMOS, a coalition of opposition groups.

Slovenia declared its independence from Yugoslavia in June of 1991. Warren Zimmerman, who was Ambassador to Yugoslavia from 1989 to 1992, found that "[t]he Slovene's separatist nationalism was unique in Yugoslavia--it had no victims and no enemies...Their virtue was democracy and their vice was selfishness (Zimmermann, 1995, p. 7)." A brief "war" precipitated by the Serb-led Yugoslavian army ended in the latter's withdrawal as democratic reforms proceeded in the new Republic of Slovenia.

The center-left Liberal Democrats received 27% of the vote in the November 1996 elections. They took 25 seats in the 90-seat parliament. Liberal Democrat Prime Minister Janez Drnovsek faced competition from the right-leaning Social Democrats, Janez Jansa, a former Communist. The latter's party received 16% vote and 16 seats. Right-wing voters had shifted support from the Liberal Democrats coalition partners, the Christian Democrats, to the Slovene People's Party, which represented farmers. The Drnovsek coalition failed to achieve support, leaving the leadership of the government in doubt in the early months of 1997.

17.1.1 WESTERN ORIENTATION

The independence of Slovenia was recognized by the EU in January 1992 as Germany led the way toward European acceptance of the newly independent states. The United States, Russia, Austria, and other countries soon recognized the independence of both Slovenia and Croatia. Slovenia became a member of the UN in May 1992 and of the IMF, World Bank, and European Council in 1993. Slovenia seeks EU entry as well as to align with the WEU and NATO. Italy for a time vetoed negotiations between Slovenia and the EU for the former's acceptance into the Union in May 1994 because of unresolved issues related to the confiscation of properties previously owned by Italians in the coastal area of Slovenia but has since endorsed the country's EU membership.

17.1.2. CONSTITUTIONAL AND INSTITUTIONAL CHANGES

In December 1991 Slovenia adopted a new constitution, based on western models. The 90-member State Assembly, whose members are elected by proportional representation to four-year terms, has exclusive authority to enact legislation. The 40-member State Council has no legislative powers, but it may propose laws to the Assembly or ask it to reconsider its vote on any particular measure. Two seats in the Assembly are reserved for Italian and Hungarian minorities. All judges are also appointed by the assembly.

The assembly elects a prime minister, who heads the government. The president of the republic, whose office is largely ceremonial, is elected by direct popular vote for a five-year term. All citizens aged 18 or older have the right to vote. Local government is organized into 147 communities with functions related to medical and social services, and the like, which are coordinated by national ministries. It has been said that there are too many "communist ideas" remaining among Slovenian leaders. There is still no compensation for the approximately 15,000 citizens who had been imprisoned for political offenses by the Slovenian Communist Party.

The Civil Society, Political Parties

When the first free elections since 1927 were held in April 1990, 1,400 candidates from 17 parties ran for 240 seats in the parliament. By late 1991, the DEMOS coalition began to break up, and in April 1992 it was replaced by the Drnovsek government. The second election, held in December 1992, drew 85% of the eligible voters. A total of 1,500 candidates representing 25 parties competed for 90 seats in the state assembly. Milan Kucan, the incumbent president, who had been a communist party leader but who ran as an independent, received 63.9% of the vote in the direct popular contest for president.

In the parliamentary elections a complex system of proportional voting did not change the party balance appreciably. The Liberal Democrats, who have their roots in the communist youth movement, gained seats and formed a coalition with the Liberal Democratic Party, the Christian Democratic Party and a former communist group calling itself the Social Democrats. Opposition parties, the Democrats, the Green Party and the Slovene National Party lost strength in the second election.

A new law effective in October 1994 prompted smaller parties to integrate with the larger. A new large center party, the Liberal Democracy of Slovenia, was formed. The so-called right wing majority coalition may become two parties, the Christian Democrats and the Liberal Democrats, are under way. Slovenia is a small, compact country with a population of less than two million there are often close personal relations among the intellectual and power elite, with no true division between left, right and center. Essentially all parties favor an effective transition to full political independence, a market economy, privatization and a multi-party democratic form of government (Mencinger and Zuppan, 1995, p. 225 ff).

The most important political parties are the Liberal Democratic party, whose leader is the current prime minister, Janez Drnovsek; the right-wing Social Democratic Party, Christian Democrats; the United List of Social Democrats (former Communists), and the Slovene National Party and Seniors.

Media, Interest Groups

There have been some attempts to develop politically independent newspapers but most of the more important are based on particular political viewpoints. The market for independent media is said to be very small and to face many competitors from abroad (Mencinger and Zuppan, 1995, p. 228). Recent legislation has established curbs on media monopolies.

Constituent groups, such as labor unions, are gradually adjusting to civil society concepts. Those which strive to represent the needs of constituents in the democratic process now include the smaller farm units and environmental organizations. Non-governmental green organizations active in pre-transition years became politicized during the first multiparty elections. Some members have reorganized as interest groups and may be said to be have emerged as part of the civil society.

17.2. Economic Policies

Among the six constituent republics of the Socialist Federated Republic of Yugoslavia, Slovenia had the strongest economy and the most highly developed industrial sector. It nevertheless shared an economic slowdown with other EE countries in the 1980s. Reforms attempted during this period were minimally successful. When a reform government was elected in the spring of 1990, Slovenia began to move toward a free enterprise system and to encourage foreign investment. The new republic's economy has been adversely affected by the armed conflicts that followed the breakup of Yugoslavia, especially by a drop in its trade with Croatia and Serbia, and by the influx of refugees from Bosnia.

17.2.1. MEASURING CHANGE

Legislation in 1990 initiated a gradual privatization of state owned firms. Additional legislation was passed in 1992. Franjo Stiblar, a professor of economics at Ljubljana University, reports that "spontaneous privatization" has taken place despite the slow passage of formal legislation. Laws have permitted foreign investments and provided for product standardization and tax and other reform (Stiblar, 1994, p. 601; see also Mencinger and Suppan, 1995, p. 234 ff).

A proper legal framework for supportive monetary and fiscal policy has been established but overall restructuring has been disorganized; and varies in performance. Delay in the restructuring of enterprises has in turn delayed the reorganization of the banking sector. It is possible that complete financial reform may create a very large public debt, perhaps as large as one-third of the country's GDP (Stiblar, 1994, p. 611). Slovenia experienced a depression in the early nineties, as its economic indicators show. By 1994 its increase in real GDP growth went from negative to positive, and there was a slight decrease in unemployment and reduced inflation. Wage levels in

Table 17.1 Economic, Social Indicators for the Danube River Basin Transition Countries, 1990 - 1995, Slovenia

INDICATOR		1990	1991	1992	1993	1994	1995
Social Indicators							
Population millions		2.0	2.0	2.0	2.0	2.0	2.0
Population growth rate %		0.6	0.6	0.2	0.2	0.2	0.2
Infant mortality rate							
per 1000 live births		8.4	8.2	8.9	NA	8.1	7.9
Life expectancy	(M)	70.0	70.0	70.0	70.1	70.5	70.9
	(F)	76.0	76.0	78.0	78.1	78.4	78.9
Literacy rate %		90.0	90.0	90.0	90.0	90.0	90.0
Economic Indicators							
Overall Economic Conditions		1990	1991	1992	1993	1994	1995
GDP in $ bn		NA	NA	10.1	11.6	15.4	21.2
Real GDP growth %		-4.7	-8.1	-5.4	1.3	5.3	4
GDP per capita in $		NA	NA	NA	7,600	8,110	NA
Industry % of GDP		38	40.8	37.6	35.4	35.1	NA
Agriculture % of GDP		4.7	4.9	4.9	4.5	4.5	NA
Unemployment rate							
(in % of labor force, yr avg.)		4.7	8.2	11.5	14.4	14.4	13.9
Retail price inflation %		549.7	117.7	201.3	32.3	19.8	12
Exchange rate (av) SLT:$		NA	27.6	81.3	113.2	128.8	119
General government balance							
in % of GDP		-0.3	2.6	0.2	0.5	-1	-0.6
Trade balance in $ bn		-0.6	-0.3	0.8	-0.2	-0.1	-0.4
Current account $ m		526	190	926	150	459	-91
External debt in $ bn		NA	1.7	2.5	2.5	2.3	3

1992 and 1993 were higher in Slovenia than in the rest of eastern Europe by about 50% resulting in some loss in international markets. A new social pact was proposed despite opposition by trade unions and was signed in April 1994. By the time of the November 1996 election per capita income had reached $10,000 which made its citizens richer than those in EU member countries, Greece and Portugal.

Despite its reduction of inflation to below 10% a year and its switch to western markets as well as a stable currency and a balanced budget, the Economist warned that "there was not enough shock in Slovenia's economic therapy." Only half of its state industry had been privatized, growth had slowed and unemployment risen to 13% (*The Economist*, November 2, 1996, p. 52).

Farming remained primarily a private function during socialism. Because of its heavily forested and mountainous terrain mass farming is difficult in Slovenia, and farms are very small. Agriculture has represented less than five per cent of the GDP during the transition period; it can be competitive only in terms of some high quality products or cattle breeding. Government subsidies are low and designed to provide some protection from foreign competition (Mencinger and Suppan, 1995, p. 242).

17.2.3. SOCIAL INDICATORS

Despite a decline in unemployment, privatization and the end of the commitment to full employment has caused massive job displacement and created a growing social burden. From 1989 to 1993 the number of older workers in the early-retirement pension system almost doubled. The social burden on the employed increased substantially at the same time. A growing private sector with small shops and joint ventures with foreign companies and the like is offering opportunities to provide employment for the displaced workers (Mencinger and Suppan, 1995, p. 244).

17.3. Environmental Issues and Policies

Slovenia is a small mountainous country bounded on the west by Italy and the Adriatic Sea, on the north by Austria, on the east by Hungary, and on the south by Croatia. While not a riparian Danube country, Slovenia is drained by major Danube tributaries, the Drava and the Sava; over half of its land area is in the Danube catchment area. The Drava sub-catchment is subdivided further by another tributary, the Mura. Of the three, only the Sava originates wholly within Slovenia; it drains 58% of the country. The Slovenian part of the Danube river basin comprises a high proportion of forests, about 52%. The population within the drainage basin is 1,604,788 (80% of the total).

17.3.1. MAJOR DANUBE ENVIRONMENTAL ISSUES

The Drava, which flows through industrial and agricultural areas, has a chain of hydroelectric power dams. There are severe problems from metal-contaminated suspended solids and sediments; and nitrites from pesticides. Ground water quality has been affected by Sava River and industrial pollution. Aquifers have been polluted by nitrates and pesticides from agricultural development (IUCN, 1994, pp. 189-190).

Water Quality
Slovenia has a sufficient supply of water for all uses. Its urban centers, including cities and settlements, or about 70% of the population, receive water supply from ground water which needs little treatment (UNDP, 1994, pp. 161ff). Supply for the hilly rural areas is presently costly and improvements in organized systems is planned.

Surface water quality is monitored and evaluated according to a multi-layered classification system from naturally pure, useable for any purpose; all right for recreation, industry, and farming but needs treatment for drinking; a level which requires pre-treatment for any other use except industry and irrigation; and water which requires supply systems.

The Drava has chronically severe problems with heavy metals, pesticides and nitrogen compounds and the entire Slovenian reach is classified as either fair or poor. The Mura is also severely polluted, mainly by organic compounds, and is also graded as Class III, or poor. Some Mura tributaries are ranked Class IV (badly polluted). The Sava deteriorates significantly downstream of Ljubljana, with high levels of heavy metal, fecal coliform bacteria and organic compound pollution, and is ranked Class III or IV. Ground water and springs used for drinking water are polluted by agro-chemicals and drainage from land fills (UNDP, 1994, p. 163-164). Over 50% of industrial and municipal discharges receives no treatment. Data collection reflecting the effectiveness of wastewater treatment is under way.

Other Resources

The Slovenian river network hosts a large number of biological resources; fish biodiversity is particularly rich. The main threats to the fish community come from river regulation, habitat destruction, water pollution, competition with introduced species and overfishing. Wetlands and ponds are significant in all the three main river systems, but hydrological engineering structures are detrimental to their ecological and environmental integrity. One of the biological riches is the forest resource, which covers approximately 52% of the Danube river basin in the country (IUCN, 1994, p. 191).

The Strategic Action Plan recommends rehabilitation of five sites. These include the establishment of a regional landscape park which would implement a plan for the elimination of nutrients and an evaluation of monitoring systems. There are plans for more nature reserves and for an international park across the Croatian and Hungarian boundaries.

Almost one-third of the Danube catchment area in Slovenia is endangered by erosion. Sediments do not accumulate in the rivers except upstream of dams because of the swift flow of most of the watercourses in Slovenia. The sediment retention will probably increase if several dams planned for the Sava to increase electricity production are constructed.

Thermal power plants burning low-quality, high sulfur lignite coal supply about one-third of electric power and are a serious source of pollution. Smaller central heating installations burning lignite are also a problem. Several cities, including the capitol, Ljubljana, have high levels of carbon dioxide and smoke concentrations. De-sulfurization equipment is being installed with help from Austrian sources to control extensive transboundary pollution. Automobile and truck traffic across the German and Croatian borders and in Ljubljana is another source of air pollution. High concentrations of agro-chemicals have been found in soils as well as heavy metal pollution and are a threat to the quality of the water supply (Wagner, 1995, p. 103).

Pollution Sources

Increased trade with EU countries has been the occasion for an upgrading of the enforcement of emission standards at large plants but smaller production units tend to disregard such safeguards. Hydroelectric plants on the Drava, Sava, and Soca rivers are disregard such safeguards. Hydroelectric plants on the Drava, Sava, and Soca rivers are responsible for pollution as well as power facilities fueled by brown coal elsewhere. Mining also causes pollution to surface and groundwater. Industry as a whole contributes to 60% of the toxic substances pollution, is responsible for 80% of the hazardous wastes, and produces 55% of sewerage. The cooling system of the Krsko nuclear power plant should receive special monitoring as it discharges its water directly into the Sava.

Road and railway transportation of hazardous materials, which has resulted in several accidents, including oil spillages, is a serious pollution threat to surface and subsurface waters. Road and railway traffic will probably to increase since Slovenia provides the shortest gateway to the sea for land-locked countries such as Austria and Hungary.

About one-third of the population is connected to sewer systems. Several large municipal wastewater treatment plants have been built in recent years, and more are planned so that they will have a capacity of 2,067,000 equivalent inhabitants. Non-point source pollution from agro-chemicals have caused problems with ground water resources. "Apparently there are no coordinated efforts to address pesticide and fertilizer overuse (Wagner, 1995, p. 104)."

Hazardous Waste

Most waste is deposited in controlled landfills, except in the Sava catchment where about one third of the deposits are uncontrolled. Most landfills, particularly within the Danube catchment area, are probably generated and contained in treated units. Despite almost 15 years' operation of the Krsko nuclear power plant (near the Croatian border on the Sava), the storage location of low and medium radioactive waste has still not been decided. All radioactive wastes is kept in special containers within the plant itself.

Health and Hot Spots

Illness from drinking contaminated water has increased in the last few years. In 1992, for example, there were 540 cases of Giardia Lamblia, a highly contagious intestinal infection, in one water supply system. The level of nitrates in various supply sources has been found to exceed acceptable limits and to be associated with outbreaks of methemoglobinemia.

The Strategic Action Plan identifies thirteen areas in need of attention, including nine which are considered to be high priority. Nine projects are intended to replace or construct waste water treatment plants; eight of these are high priority. A municipal WWTP at the second largest city of Maribor is estimated to cost $37 million. The remainder of the projects are to treat hazardous waste sites.

17.3.2. GOVERNMENT POLICIES

The Slovenian constitution (adopted in 1991) contains specific provisions for protection of the environment. The right of the individual to enjoy a protected environment and to have information about it is provided as well as penalties for damage to the environment. The basic principles of polluter pays, integration of the environment into other policies have also been included. A new Environmental Protection Law was enacted in June 1993, establishing a number of EU environmental principles including the use of BAT (best available technology), the environmental impact assessment, and, among others, an environmental fund to provide low-interest loans for investments contributing to natural resource protection.

A new Water Act which will revise the Act of 1981 will approximate EC Council Directives on drinking water quality. It will seek to control the discharge of substances which may be threatening to people, animals, or plants. Currently Slovenia has international agreements with border states to provide for the sharing of transboundary monitoring data (Brilly et al, 1996). The Environmental Protection Act has resulted in the drafting of a National Environmental Action Plan but has not yet been able to encourage investment in water protection projects.

Key Institutions

The Ministry of Environment and Physical Planning and the Ministry of Health administer legislation affecting water supply. The former has three water-related units: the Agency for Environmental Protection and Water Administration; the Institution for Environmental Protection and Water Regime, and the Hydrometeorological Institute which is responsible for water monitoring. An "Eco-Fund" was initiated in 1994 and serves as a financial institution to provide loans for environmental protection investments at low rates of interest. An Environmental Protection Council was established the same year to oversee the effectiveness of environmental protection policies (Brilly et al, 1996).

17.3.3. TRANSITION TRENDS

Slovenia's preoccupation with the development of a market economy has caused a shift away from support for quality of life values. Its budget for environmental protection has declined for three successive years. In 1993 the government included environmental needs in its strategy for economic development, integrating the principles of the 1993 Environment Protection Act. The environmental NGO movement did not begin to grow in diversity and strength until 1994 and has not yet become a key factor in the formation of environmental policy. On the positive side the predominantly export-oriented industrial sector recognizes that the "sensitive markets and regulations of the EU countries exert strong pressure to change attitudes and undertake environmental improvements (Wagner, 1995, p. 104)."

CHAPTER 18

Moldova

18.1. Background on Reform

The new republic of Moldova has a long and complex history. Most of its present borders were within Bessarabia, which had been part of the Russian Empire until 1918 when it was annexed to Romania. The Soviet Union, claiming the territory under the 1939 agreement with Nazi Germany, assumed control of the part of Romania known as Moldova and joined it with a strip of land in Ukraine called Transdniestr. In 1944 it became the Moldovan Soviet Socialist Republic.

Following World War II it fell victim to the Moscow policy of favoring a small Russian-speaking minority over the majority Moldovan-Romanian population. Of a total population of 4.4 million, 64% are Romanian-Moldovan, 13% Russian and 13.8% Ukrainian. Stalin initiated many steps to give greater power to the latter two ethnic groups. He has been accused of fostering a series of famines which cost the lives of close to one million Romanians between 1940 and 1964 (Dinu and Rowntree, 1995, p. 194). Conflicts engendered during the domination of the Soviets continue in this small republic which is bordered on the north, east and south by Ukraine and on the west by Romania.

Moldova is the second smallest former Soviet republic and has a land area of 33,840 km^2. The Prut River which forms the Romanian boundary is a major Danube tributary. At its southern end a small strip of the country borders the Danube. Moldova declared its independence in August 1991. The current president, Petru Lucinschi defeated Mircea Snegur, who had served as president since his election in 1992, in a run-off election in December 1996.

18.1.1. ETHNIC TENSIONS

Since the end of the Soviet Union ethnic tensions have increased between the Romanian-Moldovans and the Russian and Ukrainian populations favored during the USSR period. There have been periods of violence as the Russian-Ukrainian minority form a new state, the "Soviet Socialist Moldovan Republic of Transdniestria," (Dniestr Republic) and secede. The Transdniestr region broke away in 1990 and hundreds were killed in inter-ethnic fighting in 1992. Transdniestria has refused economic reform and has received no foreign aid. A Moldovan-Russian agreement of October 1994 provided for the phased withdrawal of Russia's 14th Army from Transdniestria, where it had been supporting Transdniestrian separatists. The Russian parliament declared Transdniestria a sphere of Russian influence and the troops have not been removed. An agreement to permit the Russians to keep its military establishment in Transdniestria while allowing Moldova to reassert its sovereignty over the breakaway republic has been proposed (*Economist*, October 5, 1996, p. 51).

18.1.2. WESTERN ORIENTATION

Moldova became a member of the Council on Security and Cooperation in January 1992 and of the UN in March 1992. It is also a member of the NATO Partnership program. The EU Tacis (Technical Aid to the Commonwealth of Independent States) program has provided assistance and it is said by international agency specialists to be making head way toward greater political and economic stability.

18.1.3. CONSTITUTIONAL AND INSTITUTIONAL CHANGES

A new constitution was adopted in 1994. It provides for multi-party government, an elected president with limited powers, and a parliament. Political reform has been limited. Moderate reformers, led by then President Mircea Snegur, retained political control despite challenges from Romanian nationalists and anti-reform Communists.

In June of 1995 Snegur broke with the centrist Agrarian Democratic party to form his own Party of Revival and Concord. According to expert Vladimir Socor, Jamestown Foundation, he wanted to abandon the Agrarians to alter the 1994 constitution to shift power from parliament and the prime minister to the presidency. He was, Socor has said, afraid that the Agrarians might nominate someone beside himself for the presidential elections. Snegur tried to win support among the Romanian nationalist party to change the constitution so that the official language would be called Romanian and not Moldovan. Minorities, including Russians, Ukrainians and the Turic Gagauz, believed that this was a clandestine move to unite Moldova with Romania (Rupert, *Washington Post*, May 10, 1996, A29).

Snegur's new party deprived Agrarians of their majority and generally weakened the country's ability to reform its political and economic systems. The new President, Petru Lucinschi, formerly the chairman of parliament, defeated Snegur with 54% of the vote in a run-off election in early December 1996. The Prime Minister, Andrei Sangheli, and the new president both claimed that Snegur has caused great damage to the country.

18.2. Economic Policies

Moldova's economy is primarily agricultural, aside from some heavy industry in Balti and Chisinau, its capitol. Its industrial development has essentially been tied to the agricultural sector. Soviet authorities (1944-91) planned and built large-scale agro-industrial units engaged in crop cultivation, animal husbandry, and associated canning, milling, meat and hide processing, distilling, and brewing industries. It was also a center for military electronics, particularly missile guidance system. Moldovan wines are of high quality and a major export. One third of Moldova's industrial production and 80% of its energy capacity are controlled from the Dniestr region, enhancing concerns about its eastern border (Dinu and Rowntree, 1995, p. 199).

Table 18.1. Economic, Social Indicators for the Danube River Basin Transition Countries, 1990-1995, Moldova

INDICATOR		1990	1991	1992	1993	1994	1995
Social Indicators							
Population million		4.4	4.4	4.4	4.4	4.4	4.4
Infant mortality rate per 1000 live births		NA	NA	35.0	30.8	30.3	29.8
Life expectancy	(M)	NA	NA	64.0	64.5	64.7	64.8
	(F)	NA	NA	71.0	71.5	71.7	71.8
Literacy rate %		96	96	96	96	96	96
Economic Indicators							
Overall Economic Conditions		1990	1991	1992	1993	1994	1995
GDP in $ bn		NA	NA	n/a	16.3	11.9	NA
Real GDP growth %		-1.5	-18	-29.1	-8.7	-22	-3
GDP per capita in $		NA	NA	NA	3,650	2,670	NA
Industry,mining (NMP) % of GDP		NA	NA	33	42	NA	NA
Agriculture, fisheries (NMP) % of GDP		NA	NA	43	42	NA	NA
Unemployment rate		NA	NA	NA	NA	NA	NA
Consumer price inflation %		4.2	98	1,276	837	108	23.8
Exchange rate (av) Lei:$		NA	NA	NA	NA	4.24	4.5
State budget balance in % GDP		3	0	-23.4	-8.8	-8	4
Trade balance in $ millions		NA	NA	-37	-180	-100	NA
Current account $ millions		NA	NA	-38	176	-180	-185
External debt in $ millions		NA	650	NA	325	300	NA

18.2.1. INSTITUTIONAL PROGRESS

In April 1996 the *Washington Post* reported (Stephenson, April 14, 1996, H2) that the IMF and Moldova had reached preliminary agreement on a $200 million three-year loan; Snegur was said to have curbed inflation, introduced leu currency (Romanian rather than rubles), privatized state-owned enterprises and begun a free market with well supplied shops. The U. S. government, the World Bank and the Council of Europe have called reform efforts a model for ex-Communist states (Rupert, *Washington Post,* May 10, 1996, A29). They approved the initial moves of President Snegur which slashed government budgets and privatized a number of state-run enterprises.

The agrarian party reform package halted economy shrinkage but caused great social disruption. In February 1996 the government was $65 million behind on paying wages and pensions. A proposal that Moldova pay this amount by printing unbacked currency risked crippling the reform program. A poll in February and March showed 70% think the economy is developing the wrong way and 55% say that the country's present political condition is anarchy. In June the president gave a 90-minute television address threatening to oust the cabinet.

18.2.2. MEASURING CHANGE

The economic situation in Moldova remains precarious despite its reform program. Its economic output has fallen by 61% since 1991. The country owes Russia about $330 million for natural gas. Most of its heavy industry, including the modern Rybnitsa steel plant which is in the Slav-dominated breakaway area, has been lost. While economists applaud its privatization effort, much has to be done to rebuild the former Soviet arms and heavy industrial plants. Inflation is down to 24% in 1995 from 116% in 1994 but foreign direct investment had totalled a meager $50 million since 1991. The average monthly wage of $33 buys about 44% of a basic consumer goods (*Economist*, October 5, 1996, p. 52).

The future prosperity of the country may rely on its rich black soil, and a revival of agricultural processing industries such as wine, textiles, and tobacco. Moldova's land is intensively cultivated and grazed, and very little of the original steppe vegetation remains. Except for its productive land, Moldova is resource poor, with only modest deposits of lignite, phosphorites, and gypsum.

18.3. Environmental Issues and Policies

Intensive agricultural development during the Soviet years has had the greatest impact on Moldovan natural resources. Its soil is now more than 95% under active development. Industrial development has also contributed to air and water pollution.

Environmental policies resembled those of the Soviet states and the eastern bloc countries of the Danube. Legislation affirmed the goal of environmental protection but multiple standards for a large number of parameters were not seriously enforced.

Much of Moldova's agricultural produce was exported to the Soviet Union at very low prices without regard for soil protection. Erosion in large areas of the country (74% of the country's land area is under cultivation) became a serious problem. Surface mining as well has destroyed large areas of arable land. Ground-water resources are contaminated and increasingly scarce as a result. The legacy of the past presents problems for the Dniestr Basin and the Black Sea as well as for the area of the country in the Danube Basin (29%). A serious environmental situation is complicated by political instability which has inhibited full cooperation in the development and sharing of air and water quality data.

18.3.1. MAJOR DANUBE ENVIRONMENTAL ISSUES

Water Resources

Moldova has difficulty in maintaining adequate supplies of potable water for its population. Only 8% of Moldova's water comes from aquifers and perhaps 50% of this meets the standards formerly set by the Soviet Union. Wells near major cities are drying up and rural wells are often so contaminated with high levels of minerals and bacteria that they are unfit for use. The Prut sub-catchment in Moldova is poor in surface water resources as its tributaries experience reduced summer flows. Although total annual water resources appear adequate, increasingly there are local and seasonal shortages when demand for irrigation water exceeds supply. Droughts have recently become more frequent, which makes improved water management a priority concern for government policy makers.

The Moldova National Environmental Action Plan (NEAP) puts a high priority on a rural water supply investment program, which is expected to be accomplished in one year. It also calls for an assessment of water quality and quantity (both surface and ground) at targeted locations with improved monitoring equipment to take place in one year. Also anticipated are the development and implementation of a water sector strategy to be completed in three years. The development of a network of at least 15 large water supply systems, 10 of which should be located in the Prut sub-basin, has also been recommended. Other recommendations include the improvement of the irrigation system to counter losses through inefficient transmission networks, and a cleanup program in the northwestern part of the Prut subcatchment (IUCN, 1994, p. 156).

A marked deterioration in drinking water quality has occurred in the last ten years. The percentage of water exceeding toxic levels is high. For example, in 1993 over twenty percent of drinking water supplies failed fecal coliform tests, "a doubling from 1992, and a major factor in an increase of hepatitis cases." Monitoring data from natural springs and fountains show that 69 per cent contained nitrates and ammonium nitrogen. Twenty three percent of all samples exceeded all biological standards (Moldova, 1995, p. 6). Land is traditionally cultivated close to the edge of river banks. As a result river waters are polluted by fertilizers and pesticide residues.

Contamination results from the direct runoff of used irrigation water which contains chemicals and salts. Ground water is also polluted by nitrates. Monitoring leaves no doubt that the Prut is a major contributor to the loading of the Danube with significant quantities of nitrogen and phosphorus. Since 85% of irrigation water comes directly from the Danube, both microbiological and chemical contamination is spread to crops and to the food system.

Other Resources

Moldova was rich in biodiversity, reflected in the number of plant and animal species and a significant number of endemic and endangered species that can be found in its natural ecosystems. "Economic development and exploitation have led to a widespread reduction in the country's biodiversity, particularly with reference to steppe and wetland habitats..[in] 1989 the USSR State Committee on Nature Protection reported that there was no natural landscape remaining in Moldova (Mnatsakanian, 1992, p.37)." The cost of degraded forest ecosystems, potential economic losses due to forest infestation and poor management, and their impact on recreational and biodiversity values and watershed integrity are judged to be high but difficult to measure because of a lack of data (Moldova, 1995, p. vi).

The IUCN report identifies three nature reserves established recently: the Criva Nature Reserve, a Karstic complex in the northwestern sub-catchment; the King Forest - a virgin deciduous forest; and Lake Baal, an important wetland located within the floodplain of the lower Prut river. The evaluation of the existing network of protected areas in preparation for a management plan is under way. The plan is to extend the network of protected areas to cover the entire floodplain of the Prut (23,000 ha), a highly important wetland area (IUCN, 1994, p. 157). The Strategic Action Plan recommends that 250 km of the lower Prut corridor, a protected Ramsar site with two national parks, be restored with respect to its hydrological regime, the extension of its alluvial forests, and the improvement of its general management.

Agricultural, Air Pollution

Land use practices have not only resulted in extensive top soil loss and reservoir siltation but have also affected water quality, soil fertility, and the life-cycle of downstream artificial impoundments. Deposits of mud layers as deep as six meters have been recorded in most of the 21 reservoirs. Recent investigations show high levels of particulate organic matter and significant levels of heavy metal and pesticide accumulation in reservoir sediments (IUCN, 1994, p. 157).

Air pollution, primarily caused by vehicles, is common in large towns. It is also caused by stationary sources. The giant regional coal-burning plant near Dniestrovsk accounts for 53% of all air pollution from stationary sources. The increase resulted from demand for greater output following the shutdown of Chernobyl. Thermopower plants, small industrial enterprises, and large open deposits of organic wastes produced by very large animal husbandry complexes have also been identified as the main air pollution sources. Explosions and other accidents in the arms plants have also caused serious air pollution problems. Soil contamination with agro-chemicals has also affected the food supply, but data about levels are not adequate for research purposes.

Pollution Sources

The industrial sector contributes to half of the gross national product; pollution sources of major concern are the food industry, primarily sugar factories, poultry farms and dairies. Discharges from these sources and from other industrial enterprises do not receive adequate treatment before discharge. A large number are connected to overloaded municipal treatment plants. While a number of industrial wastewater plants have been constructed many industries either do not use their facilities or fail to maintain them in good condition. Biological sludge from plants is frequently high in heavy metals and unfit for agricultural use.

Many agro-industrial plants also contribute to pollution in the same way. Industrial and other activities have induced changes in the hydrology of the Prut in recent years where poor water management has deprived a large farming area of floodwater recharge. Erosion and extensive gravel exploitation have lowered the bed of the Prut and caused significant damage to the hydrology of the region.

Domestic sewerage and waste water are also major sources of surface and ground water contamination. A significant proportion of the population is not connected to sewer systems. Most of the villages in the country side discharge wastewater without treatment. Many of the treatment plants which handle an estimated 70% of the population are overloaded. Since the volume of domestic sewerage and wastewater is not affected by the economic downturn which has decreased industrial pollution, the pollution load has increased in recent years.

The wastewater produced by agricultural processing plants is loaded with organic matter discharged along with domestic waste into the municipal sewer system. Priorities for agriculture include the need to improve and extend irrigation to deal with drought and to minimize the use of fertilizers and pesticides. Intensive agriculture, including both crop and animal production, continues to result in high levels of pollution. There are a large number of small livestock farming enterprises which contribute to environmental degradation, polluting both air and water with nitrates and ammonia. Many pesticides have been banned but remain in storage; others still have passed their "use by" date. The storage of such materials is extremely hazardous since they lack proper management.

Landfills
Thirty per cent or more of landfills currently exceed capacity; improvements have been stopped because of the current economic crisis. A significant amount of organic wastes is annually produced by the large animal complexes and these are frequently disposed of in the vicinity of landfills. These huge organic deposits have been identified as the main sources of groundwater pollution. Solid wastes are disposed of in a large number of unsecured and uncontrolled small landfills. Their impact has not yet been studied.

Health and Hot Spots
Ground water quality is deteriorating because of leaching form intensive pig farming, among other agricultural activities. Ground water protection zones exist in name only. A spokesman for the Moldovan Institute of Hygiene and Epidemiology has advised that from a public health point of view ground water should no longer be used for drinking water supplies (UNDP, 1994, p. 96). The impact of contamination on the population is serious. Infant mortality and mental retardation rates are unusually high, reflecting the fact that more than 70% of the population is not supplied with treated drinking water. Drinking water is taken from shallow wells; data show very high nitrate concentrations.

The Strategic Action Plan has identified seven areas, all high priority, in need of corrective action. These include three municipal waste water treatment plants, and corrective action on three lakes and a pig farm. Moldova's NEAP recommends a number of priority measures which include the disconnection of industrial effluent containing heavy metals from municipal wastewater treatment systems and enforcement of pre-treatment standards. These measures are to be followed by a capital maintenance investment program for a municipal water and wastewater treatment infrastructure. An inventory of pollution sources in residential settlements are expected to result in a set of local programs to eliminate sources of household pollution and improve the sanitation of their surroundings.

18.3.2. GOVERNMENT POLICIES

Most legislation now in force is a product of the past. The State Department for Environment and Natural Resource Protection created in 1990 is presently re-drafting water protection and other legislation. The use of pesticides and fertilizers is controlled by laws issued by the Department for Environmental Protection. Framework legislation was introduced in 1993 (the Law of Environmental Protection), together with laws dealing with specific areas of environmental protection, forests, water and mineral resources. Others to be adopted will apply to air quality; protection of flora and fauna; waste and hazardous substances; protected areas; and natural resource use taxes (IUCN, 1994, p. 160).

New land laws establish the general basis for use of environmental resources and introduce a system of charges for pollution which is within permissible limits, make environmental impact assessment obligatory, establish the principle of biodiversity maintenance and also specify the rights and responsibilities of industries which exploit the environment. Moldova has also begun to introduce economic incentives for environmental protection by means of a pricing policy, including user charges, taxes and subsidies. The charges for water use are now required to cover the total cost of water supply and water protection. Pollution charges paid by industrial sites discharging waste water generally depend on such factors as the quantity of waste water, concentration of polluting substances and the quality of surface water. Compensation is payable for the damage of natural resources, including air, fisheries, water, soil, and forests.

Extra-budgetary environmental funds have been created at the national, district, regional and municipal level to supplement existing funding sources for environmental protection. Sources for the funds include land tax, water use taxes, pollution and illegal resource exploitation fines and voluntary donations. The uses to which the funds may be put are restricted to projects relating to environmental protection.

Key Institutions

The State Department of Environmental Protection is the main policy-making agency for all environment related matters. It was created by the parliament, to which it reports, in July 1990. Ministerial status for the State Department of Environmental Protection has been recommended as well as other organizational changes. The Department prepares an inventory of natural resources and standards in relation to emissions, waste storage and natural resource use, and legislatioon.

18.3.3. POLICY IN TRANSITION

Moldova's NEAP, developed in collaboration with the World Bank, sets goals for changes in economic policies as well as improvement in the protection of the environment.

The NEAP incudes:

- changes in economic and fiscal policy which would phase out subsidies to energy, water, raw materials and agricultural inputs and resource consuming enterprises;
- phased increases in user fees for water supply, wastewater treatment, hazardous waste collection with full costs eventually covered through the user pays and polluter pays principles;
- an environmental fund as a self-financing mechanism;
- strengthening the institutional roles of administrative units for enforcement purposes including monitoring and coordination among government units;
- adopting a full agenda of environmental laws and regulations;
- encouraging the role of the private sector to increase efficiency;
- initiating programs in each important environmental issue area including public awareness.

Time schedules for target accomplishments were set over a one, two, or three-year period. The targets have already been included in new legislation.

CHAPTER 19

Ukraine

19.1. Legacies

Ukraine, the second largest of the former Soviet republics, suffered severely in the aftermath of World War I. The eastern part of what had historically been called Ukraine was conquered by the Soviets in 1921 while its western land area was annexed by Poland, Romania, and Czechoslovakia. Stalin's farm collectivization policies caused extreme hardships for the country. An artificially induced famine in 1932-33 killed an estimated seven million people. "All food stocks were forcibly requisitioned; a military cordon prevented all supplies from entering; and the people were left to die. The aim was to kill Ukrainian nationhood, and with it the 'class enemy' (Davies, 1996, p. 965)." When the German armies invaded in 1941 many Ukrainians hailed them as liberators.

World War II brought massive destruction. Postwar Soviet annexation of western Ukraine was resisted by guerrilla forces until the early 1950s. Crimea was returned to Ukraine in 1954 but it is still causing problems as the majority Russian population continue to support its return to Russia.

Russification

Russification was practiced under Communist party chief Volodymyr Shcherbytsky. Of the 51.7 million inhabitants of Ukraine, 72% are Ukrainian; Russians are the largest of several minority groups, with 22% of the population. The People's Movement of Ukraine, the Rukh, sought greater autonomy in the 1980s. It gathered momentum; independence was proclaimed in August 1991 and confirmed in a referendum in December. A former Communist, Leonid Kravchuk, was elected president, defeating the Rukh candidate, V. M. Chornovil, with 62% of the vote to Chornovil's 23%. In 1994 Kravchuk defeated by pro-Russian former prime minister, Leonid Kuchma, the head of Ukraine's biggest rocket factory, who won 52 percent of the vote.

Western Orientation

Ukraine's policies toward the west have been positive, but only to a degree. The country is part of the NATO Partnership Program and has participated in military exercises with US and European troops. Along with Russia it joined the Council of Europe last year. The surrender of its nuclear arsenal was completed in June 1996 with the help of funding from the US. It has been a member of the UN since 1945.

19.2. Political Reform

19.2.1. CONSTITUTIONAL AND INSTITUTIONAL CHANGES

Ukraine still uses its substantially amended 1978 constitution. The draft of a new constitution after independence remained a matter of discussion through 1996. Its terms were weakened when private property ownership was excluded as a basic right. Many provisions of the 1978 constitution were superseded by a special agreement reached between President Kuchma and parliament in June 1995.

In the summer of 1991 parliament voted to create the post of president, and the first presidential election was held in December. Executive power is exercised by a council (cabinet) of ministers headed by the premier, who is appointed by the president subject to parliamentary approval.

The Civil Society

When parliamentary elections were held in the spring and summer of 1994, the communists and their allies won the largest bloc of declared seats, but an even larger number of seats went to candidates with no declared allegiance to a party. Major parties in parliament are grouped in political factions and groups rather than parties per se. The communists in 1996 had 88 out of 450 seats; the Ukrainian Peoples Movement for Restructuring 29, and the socialists, 28; reformers held 31 seats. Ukraine was described in the *New York Times* of June 27, 1996, as "bedeviled by a corrupt and backward-looking political elite dominated by former Communists and socialists."

Media, NGOs

A nascent civil society, a free media and active, representative, nongovernmental organizations, has not as yet shown strength. Parliament voted to investigate the Ukrainian broadcast media following the report of a temporary commission supported by left-wing members. The decree proposed that the invasion of "foreign capital" into radio stations be investigated. The president also has a commission to investigate media.

19.3. Economic Policies

Ukrainian agriculture, traditionally important, remains collectivized. Central planning, directed by the Soviets in the post-World War II period, provided for increasing industrialization, primarily in eastern Ukraine. Soviet policies, as in the eastern bloc countries, favored heavy industry over light and lower production of consumer goods. Industries included basic chemicals, tractors and automobiles, aircraft and electronic instruments. The steel industry was rebuilt and expanded. The energy sector increasingly came to depend on nuclear power (Freeman, 1995, p. 151). Ukraine also specialized in the production of military equipment. The usual distortions in economic development occurred as the country became dependent on the other bloc countries for supplies and markets. The country was also heavily indebted to Russia.

Table 19.1. Economic, Social Indicators for the Danube River Basin Transition Countries, 1990 - 1995, Ukraine

INDICATOR	1990	1991	1992	1993	1994	1995
Social Indicators						
Population million	51.9	51.9	52.2	52.2	51.9	51.4
Infant mortality rate per 1000 live births	13.0	14.0	14.3	15.1	NA	NA
Life expectancy (M)	NA	NA	65.0	65.3	65.5	65.6
(F)	NA	NA	75.0	74.7	74.8	74.9
Literacy rate %	98	98	98	98	98	98
Economic Indicators						
Overall Economic Conditions	1990	1991	1992	1993	1994	1995
GDP in $ bn	NA	NA	NA	132	102	90
Real GDP growth %	-3.6	-11.9	-17	-14.2	-24.3	-11.8
GDP per capita in $	NA	NA	NA	3,960	3,650	NA
Industry, mining (NMP)						NA
% GDP	41.3	42.4	50.7	52	NA	
Agriculture, fisheries (NMP) % GDP	30.3	30.2	22.5	16	NA	NA
Unemploy. Rate (end-year)	0	0	0.3	0.4	0.4	NA
Consumer price inflation %	4.2	91	1310	4735	891	380
Exchange rate (av) Krb:$	NA	NA	241	15,415	68,000	160,000
General government balance in % of GDP	NA	-13.6	-29.3	-9.7	-9.1	-4
Trade balance in $ bn	NA	NA	NA	NA	NA	NA
Current account $ bn	NA	-2.9	-0.6	-0.8	-1.4	-1.3
External debt in $ bn	NA	10.4	12	NA	7.5	NA

19.3.1. INSTITUTIONAL PROGRESS

The transition to a market economy, supported as a goal by the new government, has been moving at a slow pace. In the absence of a constitution, legislation supporting privatization was enacted in 1992. There has been progress in the service sector but factories continue to be run by the state.

The government, lead by a weak president with a strong Communist membership in parliament, has continued to prefer collectivization to privatization. The industrial base includes a number of Chernobyl-type power plants, steel works, armament factories, engine plants, and satellite research sites. Indications are that more than half of the economy is run on the black market, the result of high taxes levied on businesses (as high as 89%). Because of this fact and despite the lure of 52 million potential customers, foreign investment is very low, $700 million by 1996, compared to Hungary's $13 billion.

19.3.2. MEASURING CHANGE

Ukraine has received foreign aid from the U.S., the IMF, the European Union Council, and Japan. The U. S. AID has loaned over $ one billion, putting Ukraine ahead of Russia on its list of recipients and behind only Israel and Egypt. There were 52 AID projects in 1996. Despite such aid the *New York Times* reported in June 1996 that millions of workers had not been paid in six months and that both economic and political conditions were close to chaos (Perlez, *New York Times*, June 27, 1996).

A $900 million loan from IMF was withheld in May 1996 pending improvements in the promised economic reform package. The IMF has been patient with the slow pace of reform and satisfied with promises over performance (*Economist*, May 4, 1996, p. 52). The economy shrank by 12% in 1995, and although inflation was down from 10,000% in 1993 it was still 180% in 1994. The President continued to say that mass privatization would be complete by December 1996.

19.4. Environmental Issues and Policies

Ukraine, second in area after Russia and sixth in population among European countries, occupies only 4% of the total Danube catchment area. Of the 3 million people living in the area (5% of the total population), 42% reside in urban areas and 58% in rural. The Danube catchment area divides into the Transcarpathian region within the Tisza basin, the Upper Prut region which includes the Siret catchment, and the northern part of the downstream Danube Delta along the Kilijan arm of the Danube.

There are major differences among these regions with respect to geography and land and water resource use (IUCN, 1994, p. 196). The Ukrainian shoreline of the Danube measures only 174 km. This part of Ukraine experienced the full brunt of repressive Soviet policies with respect to all aspects of its political and economic life. Intensive agricultural, urban, and industrial development was accomplished with a minimum of effective protective measures. "Soviet economic development left Ukraine heavily reliant on nuclear power (30 percent), beset with industrial pollution (one of the highest per capita rates in Europe), short on water (second to last per capita among the former Soviet republics), and faced with the cleanup of the worst nuclear accident in history (Freeman, 1995, p. 160)."

19.4.1. MAJOR DANUBE ENVIRONMENTAL ISSUES

In the Danube drainage area there are three significant rivers: the Tisza with 42% of the catchment area, the Prut with 31%, and the Siret with 7%. Most of the run-off comes from the upper Tisza, an international river system which is shared with Romania, Slovakia and Hungary.

Water Resources

The UNDP Health Report states that "No surface water in the region can be drunk without treatment if compared to the national and WHO standards for drinking water quality. This is partly due to discharge of large quantities of (untreated) waste water, partly due to heavy pollution from non-point sources, mainly fertilizers and pesticides." With respect to ground water, ten years ago only ten out of 25 oblasts (districts) reported pollution from pesticides, nitrates, heavy metals and radio nuclides; in 1993 all reported contamination from these elements. Fifty per cent of the wells in the Odessa oblast are said to be contaminated with nitrates and pesticides exceeding WHO guidelines (UNDP, 1994, p.193).

The Ukrainian Ministry for Environmental Protection and Nuclear Safety publishes an annual report on the state of the environment. "The State of the Environment in Ukraine, 1994", Kiev: Rayefsky Scientific Publishers, 1996, is available in English. Its statistics support conclusions reached elsewhere.

The quality of Ukrainian water resources is evaluated according to the state standards developed by the Ministry for Health Protection. These standards have been established for 2,000 elements and compounds but have been elaborated for only 28 of them. The guidelines specify the maximum acceptable concentrations of pollutants for various uses but are clearly in need of modernization and of stricter application than they now receive.

Other Resources

Almost all of the forested land of Ukraine is located close to air pollution sources, national and transboundary. The IUCN report reviews the status of biodiversity in the Tisza, Prut and Danube catchment areas. The Danube region is a major source of fish for the Ukraine. While quantity has remained constant over the last four decades, diversity has decreased sharply with only 17 of 40 commercially important species still available. Diverse woodlands cover 50% of the Tisza region which has 400 species of trees and bushes. Timber activity has altered the forest community structures by maintaining a high percentage of sapling (40%) and middle age stands (30%) in the region. The Prut basin has a high level of diversity sustained by the Carpathian National Park. The Danube Plav is an international biosphere reserve which has a large number of vascular plant species and of vertebrates.

The Strategic Action Plan recommends projects to enlarge the existing Danube Delta Reserve in the lake area between the Danube and the Dniestr rivers, a 500 km^2 floodplain area. Another project in the Transcarpathian region of 10,000 km^2 would provide partial restoration and establishment of protected areas.

The suspended matter of the Danube, primarily fine particles such as silt and clay, comprises its main input to the Danube Delta where it accumulates. The Prut and the Siret rivers are characterized by high loads of suspended-matter, even within their upper reaches. The Tisza and its tributaries also carry out of Ukraine a large quantity of fine particles.

Air Pollution

The Chernobyl nuclear power plant, north of Kiev and on the Belarus border, was the site of the world's worst nuclear-reactor disaster on April 26, 1986, when the plant's No. 4 reactor exploded. The disaster killed 31 persons immediately or shortly thereafter and caused the hospitalization of about 500 others. According to unofficial accounts, deaths which are attributed mostly to work on cleanup are believed to have been as high as 10,000 (Freeman, 1995, p. 153). Mnatsakanian reports that despite substantial funding from abroad to finance resettlement, hundreds of thousands of people were put into the zone to build a new residential area for plant personnel and for other reasons. Army recruits were used to clean the territory immediately after the catastrophe (Mnatsakanian, 1992, p. 57).

Although heavily contaminated soil and trees were removed from the 30-km (19-mi) zone near the power plant, authorities acknowledged in 1990 that several million persons were still living on contaminated ground. The incidence of thyroid cancer, leukemia, and other radiation-related illnesses are higher than normal among this population. At the plant itself, reactor No. 4 was entombed in concrete. Two of the three remaining reactors continued in operation, but a series of accidents at Chernobyl persuaded Ukraine's parliament in 1991 to press for a complete shutdown,-an unlikely event until another source of power for the region is developed.

Each of the three regions in the Ukraine Danube catchment area has its own specific air pollution problems, resulting from particular anthropogenic activities. All have high concentrations of sulphur and nitrogen oxides from fuel combustion, both industrial and heating units, and hydrocarbon oxides and additional nitrogen oxides from public and private transportation. The soils of the three regions receive a fairly large amount of atmospheric pollutants; soils used for agriculture are often loaded with nutrients and pesticides. There has been significant soil erosion due to the use of heavy mechanical equipment.

Pollution Sources

Industries in all three regions lack facilities to pretreat discharges of waste into surface water, creating water pollution problems. The Tisza region, which has the highest percentage of industrial development, is dotted with food factories and pulp and paper mills which release toxic emissions into air and water. The main pollutants are dust, soot, carbon and nitrogen oxides and sulphur. Air pollution also comes from public and private transportation sources which lack catalytic converters. Waste from Prut area oil and sugar refineries, fertilizer and wood working plants is responsible for high levels of toxic pollutants in streams. The presence of fog over the Danube region during winter months, which can stay for 20 days at a time, contributes to a high level of air pollution (UNDP, 1994, p. 202).

The annual discharge of two cities in the Tisza region averages 87% of the total discharge of municipal wastewater in the region. Their wastewater treatment plant is severely overloaded. In the Prut region, several municipalities either lack a wastewater treatment plant or are using one that needs an extension of its capacity because of it is overloaded.

Agriculture

Agricultural methods have been a main cause of the degradation of water quality in all parts of the catchment areas. Agro-chemical products and the location of large livestock farms have been responsible for heavy contamination by nutrients and pesticides so that ground water is contaminated as well. "Regarding point source pollution [in the Tisza region], the 4 pig breeding complexes of 7-8,000 heads each, produce altogether 150,000 m^3 of liquid manure annually, contaminating soils and groundwater. Agricultural run-off from farms is estimated to be responsible [for] 70% of the overall pollution with livestock farms contributing the remaining 30% (IUCN, 1994, p. 203)." Mechanization has heavily eroded arable land in all three regions and dumpted large amounts of suspended matter into the rivers. Irrigation use of highly mineralized water has caused the salinization of the soil, limiting its arability.

Solid Waste

The main producers of solid wastes are located in the Tisza and the Danube regions. No more than 10 to 15 percent of industrial and solid waste is recycled. The rest is deposited on surface dump-sites, accumulating in sludge ponds, refuse heaps, etc. The total surface under such sites throughout Ukraine has reached 160,000 hectares, with the total waste weighing nearly 25 billion tons. Waste is a prime factor in pollution. Air emissions in particular pollute surface water, soils, and reduce arable land, and have a serious impact on human health.

Hot Spots and Health

Health authorities in Ukraine have acknowledged that there have been massive outbreaks of intestinal infections in recent years. These epidemics have come from contaminated water supply systems. One expert maintains (without official confirmation) that 80% of all diseases are caused by the low quality of the country's drinking water (UNDP, 1994, p. 195). Low levels of fluoride and iodine have been the occasion for dental caries and endemic goiters. In 1992 in the Odessa area (2.6 million inhabitants) there were 14,000 cases of enteritis, dysentery and hepatitis A. Several health authorities claim that the content of radio nuclides in drinking water due to the Chernobyl accident is a matter of concern and should be studied in greater depth than it has been. The Danube part of the country may be less affected than other parts of the country but nevertheless should be part of any such investigation.

Possible Projects

Four possible environmental projects, three of them high priority, have been identified by the Strategic Action Plan. One would modernize and improve the capcity of an existing waste water treatment plant. The other would provide support for the construction of a new one at Chernivtsi in the Prut basin. Two projects in the Tisza basin would improve the technology of two plants, a wood processing and a chemical enterprise so that their discharges meet emission limit values (EPDRB, 1995).

19.4.2. GOVERNMENT POLICIES

Legislation inherited from its days as part of the Soviet Union is still in effect in Ukraine. Framework legislation passed in 1991 still lacks the additional laws which are needed to bring environmental quality standards and their enforcement to all aspects of environmental protection. "As a result, the legal authority of the new environmental ministries remains ambiguous. At the same time, the structure and functions of new ministries remain largely unchanged (Freeman, 1995, p. 162)."

Key Institutions
The Ukrainian Ministry for Environmental Protection was established in 1991. It assumed the responsibilities of the former State Committee for Protection of the Environment. Its duties are broad and include the development and enforcement of all policies related to environmental resources. It is in charge of the national environmental monitoring and information system; organizes and controls the establishment and evaluation of standards, approves legislation, issues discharge permits, and supervises protected areas. It also ensures the fulfillment of Ukraine's international obligations. It lacks staff, particularly economists, policy analysts, and environmental managers, and is top heavy with those who have experience elsewhere, or have backgrounds in science and engineering. Its budget is low in comparison with its tasks (Freeman, 1995, p. 163).

In addition to the environment ministry there is a Ministry of Environmental Safety. This ministry supervises environmental and economic reforms which will provide economic incentives for environmental protection. There is a state committee for water resources, for hydrometeorology which monitors surface water quality, a state committee for housing and community services with some responsibilities for water supply, the ministries of health, agriculture, and geology all of which have monitoring and other crossover functions. Fragmentation of environmental responsibilities has been the occasion for criticism and for proposals to streamline the environmental protection function.

19.4.3. POLICY IN TRANSITION

Soviet policy of the late 1980s endorsed some of the major aspects of the EU council directives on the environment but they were never put into practice. The Ukrainian government has not yet adopted western European environmental standards, permitting and compliance programs, the collection of pollution fees and the development of environmental funds, environmental impact assessments and the like. Cumbersome administrative procedures and reduced budgets have seriously interfered with updating of environmental policies. The Ministry is more a technical support than an enforcement agency. These and other problems with the development of a clear, centrally organized and monitored set of regulations, remain of concern to environmentally conscious citizens and neighboring countries in western Europe.

PART FIVE: Long-term Commitments

CHAPTER 20

Agreement and Programs

20.1. Cooperation Background

Cooperation among Danube countries to protect their vital, shared natural resources was sporadic and ineffective throughout the Cold War. The basin lacked almost all of the common denominators of effective river basin management. Agreement to negotiate resource issues, in the face of obvious and widespread degradation of air, soil, and water quality, was lacking. Formal cooperation was limited to support for a commission which regulated navigation and avoided questions of water quality. Support for a basinwide resource management plan was not under discussion.

In any case, such a plan required the sharing of accurate and compatible data about pollution levels and the effectiveness of resource protection policies in each country. Most environmental data were not only incompatible and inaccurate but also inaccessible within countries and rarely shared among them. Countries in eastern Europe and the Soviet Union readily signed international conventions to protect wetlands and reduce air pollution, but signatures on documents did not bring change. Resources for active programs to protect the environment were lacking in the countries themselves and not available from third parties in west or east.

20.2. Program Accomplishments

The years from 1991 to 1996 hardly seemed ripe for the development of an international river basin management program in the Danube Basin. An historic framework for resolving basin-wide issues was lacking. Pressures on governments attempting to administer political and economic changes were immense. The unresolved conflict in the former Yugoslavia hurt the Danube countries economically and were a sorry reminder of past and possible future failures to negotiate the ethnic problems present to one degree or another in each of the basin states.

And yet, by the end of 1995, the Danube countries, supported by "third parties" concerned about them, international organizations, in particular, the EC and the members of the Global Environmental Facility (GEF), had taken great strides.

The countries and international organizations working together had

- signed a Convention on Cooperation for the Protection and Sustainable Use of the Danube River (Danube River Protection Convention);
- pending formal ratification and the entry into force of the Convention supported an interim International Commission for the Protection of the Danube (IC) and a Commission secretariat;
- developed the framework for and began construction of an extensive data-sharing network which could use compatible, state-of-the-art, monitoring and measuring techniques and permit the periodic evaluation of the water quality of surface waters throughout the basin;
- supported the strengthening of democratic institutions relevant to environmental protection issues;
- agreed on a plan which identified sites most heavily impacted by water resource pollution and recommended action to rehabilitate them and protect them in the future, and
- developed a transition format for the functioning of the International Commission secretariat.

20.2.1. STATUS OF PARTICIPATING COUNTRIES

The nine new democracies participated at different levels in the series of activities which the Danube Program Task Force laid out for Phase I (1992 to 1995). Participation was the product of a number of factors which have been summarized in the profiles in Part Four. It depended on political stability and whether or not economic and social conditions could tolerate the substantial investment required for the rehabilitation of basin natural resources and their future protection. Some countries, by the end of 1996, appeared to be considerably closer than others to such a goal.

Environmental Concern
Participation also reflected levels of environmental concern among political leaders, entrepreneurs, and the newly enfranchised electorates. Reports of damage to natural resources made public and widely shared for the first time made it possible to identify rehabilitation needs in each country. The potential cost of reaching parity with policies advocated by the EU was staggering. It appeared that only two countries, Hungary and the Czech Republic, were ready to fund new treatment facilities which would help them to meet western environmental standards.

Basis of Protection
The following chapters describe the cornerstones of long-term river basin environmental protection for the Danube: the commitments countries have made to the Convention; the basin-wide sharing of data which should make it possible to evaluate the status of water quality in the whole basin and register the effect of protection policies; the strengthening of institutions, particularly non-profit NGOs concerned about national and local policies

to protect people and the environment from pollution, and the identification and rehabilitation of the most severely contaminated sites through the Strategic Action Plan.

20.2.2. THE CONVENTION

The Danube River Protection Convention (Danube Convention) commits participating countries to the adoption of generally compatible environmental policies. In contrast with a proposed ecological convention supported by the Hungarian government, it concentrates on water quality issues but refers to the need for agreement on improved protection for all threatened resources. Chapter 22 explores the pre-transition history of agreements among basin countries and contrasts their objectives and impact with the principles embodied in EU council directives on the environment. The objectives and major activities of the IC are rooted in international conventions on the protection of water and other natural resources, particularly the UN ECE sponsored Convention on the Protection and Use of Transboundary Watercourses and International Lakes signed in Helsinki in March 1992.

The Danube Convention supports the continuation of the major activities of the Danube Program: monitoring and sharing of water quality data; the harmonization of environmental policies among all countries, and the continued support for basin-wide plans to invest in the improved protection of basin water resources. During what may be considered a transition period for the new IC plans are under way to establish complementary work groups to match the Danube Program activities.

20.2.3. AN OPERATIONAL PROGRAM

The Program Coordinating Unit of the Danube Program (Part 3) has reported in considerable detail on its activities from 1992 to 1996, including annual reports and interim summaries of the activities of sub-groups. Each of the three major activities, database networks, help to the civil society, and implementation of a strategic action plan is discussed in separate chapters. Two concluding chapters describe the future of international management in the basin: plans for how it might operate under a legal agreement and for how it might be funded in the future.

20.2.4. DATA AND INFORMATION SOURCES

The second item on the short-term work plan action agenda, the development and dissemination of an accident emergency warning system, was on its way in 1993. The sub-group in charge based its work on the examples of systems in other river basins, including the Rhine. It assigned highest priority to developing and transmitting information about the presence of dangerous substances or other emergency situations which could cause transboundary impact. Problems included the use of many different languages in the basin and the communications gap resulting from the situation in Bosnia and Serbia (Chapter 23).

Data Lacking

Aside from the national reviews requested by the Danube Program Task Force in 1992 occasional studies there was little or no data about the environmental status of the Danube countries available from 1992 to 1996. Countries issued yearbooks about the condition of their natural resources, but these were not widely available, nor were they necessarily mutually compatible. An initial inventory of data needs was conducted by a consortium of firms from the UK, Netherlands and Denmark with help from Danube country consultants. Monitoring practices were reviewed for their flexibility in meeting long term and emerging needs and to support trend analyses for use by policy makers. If data were to be shared across national boundaries, testing laboratories had to use compatible, state-of-the-art sampling methods, analytical procedures and quality control. Quick and easy access was to be provided to water quality data and to general information about Danube Program activities.

NATO Workshop

The NATO sponsored workshop held in May of 1996 brought key actors from the Danube Basin countries and the Danube Program to review and make recommendations for improving electronic access to water quality data and information about the basin (Murphy, 1996). Their conclusions and suggestions are included in Chapter 23. A follow-up project to prepare pilot systems at a central location was scheduled to be held in 1997. A NATO Linkage project was to begin a network between member and partner countries.

GEF-EU Phare Support

The Global Environmental Facility joined the EU Phare program in funding several activities, including inventories of biodiversity sites in need of protective measures, pre-investment studies, and general help to the Program Coordination Unit and the Task Force. Through its special focus on institutional strengthening it helped environmental governmental ministries and related agencies on the national and local levels. It also specialized in funding non-governmental environmental organizations (Chapter 24). Relatively modest investments have been made for the latter two programs.

20.2.5. A STRATEGIC ACTION PLAN

A strategic action plan took shape in 1993 and was adopted by all participants in the Danube Program in 1994. It identified high priority targets in 15 tributary basins and worked with governments in each country to construct an "investment pipeline" to implement the plan. The Plan was intended as a guide to enable riparian countries to coordinate the activities of local and national authorities, international financial institutions and regional and non governmental organizations to reach a common set of goals.

A biodiversity inventory, limited to the immediate river corridor and its main tributaries, provided information on a number of priority areas for habitat management, protection and rehabilitation. Specific suggestions for the implementation of the plan have been formulated and are supported by individual National Action Plans prepared by each country.

20.2.6. APPLIED RESEARCH PROGRAM

The preparation of a program of applied research will help to support the implementation of a framework that is required to deal with the concern over the environmental situation of the Danube River, the Delta and the Black Sea. These funds are earmarked to provide for specialized equipment, workshops and other activities. Some research funds may be needed to support other improvements such as in the Monitoring, Laboratory, Information Management (MLIM) activities

20.3. Transition

A transition is under way so that the IC will assume the responsibilities of the present task force and its secretariat will take over the operating tasks of the Program Coordination Unit. Long-run tasks for implementing the Plan are being identified. Representatives of the country governments, for the most part high level staff in ministries of environment, have developed expertise over the last four years, more data are available to help meet Program goals, and in other ways it seems likely that the new Commission could function well (Chapter 25). The missing ingredient is funding. Needs and possible sources are explored in Chapter 26.

CHAPTER 21

The Danube Convention And Its Commission

21.1. An Early Endorsement

The Danube Program launched and guided urgent measures to restore the damaged natural resources of the River Basin and to protect them in the future; the Danube Convention endorsed at the ministerial level in 1994 provided a long-term mechanism for strengthening and implementing the joint environmental policies thus begun. The 1994 signing of the Convention by all Danube countries except for Bosnia and Serbia was an historic occasion in more than one sense. The Danube agreement was one of the first agreements on a transboundary waterway to deal specifically with the protection of environmental quality. The basin is one of the few basins where a long period of dissent among countries has been followed by a detailed agreement on the significant basin management issue of water protection and use.

Pending ratification, funding support, and the continued good will of all the countries, it will contribute to the long-run protection and sustainable development of the Danube Basin, be a boon to the region economically, and demonstrate the wisdom and feasibility of negotiating basic conflicts over resource use.

21.1.1. EARLY RATIFICATION

The rapid advance of the Convention from draft document to ratified agreement is the result of many factors. Few questioned the need for a separate Convention, distinct from the navigation agreement administered by the Danube Commission. The long term navigation agreement reached in 1948 to establish the Danube Commission has built-in political problems. Its membership excludes Germany, providing the country only observer status, and includes Russia despite the fact that the country is not in the Danube catchment area. Its official languages continue to be Russian and French. These and other factors prevented the Belgrade agreement from extending its prerogatives to environmental protection.

Initial Proposal

The Convention proposed by the Austrian government in 1991 drew instead on the increasing willingness of the former communist countries to harmonize economic, political, and environmental policies with those of European Union countries. Other international environment treaties had attracted signatures from the eastern bloc countries during the 1980s so that the countries themselves had helped to pave the way for regional accord. The advantages of harmonized environmental policies to both sides were increasingly apparent. With the end of the Cold War basin-wide environmental management measures received further strong support from the initiatives of the UN Economic Commission for Europe and the establishment of the European Environmental Agency.

A "Rival" Convention

At the February 1991 meeting to discuss a possible environmental convention, hosted by the Hungarian Ministry of Environment, the Danube Basin Ecological Convention, was put forward by the Hungarian representatives. This Convention is still under discussion but has not been submitted for formal approval or ratification. Its objectives focus on all ecological factors, rather than water quality management alone. It also includes Switzerland, Italy, Poland, and Albania among cooperating countries. These four have been excluded from the Danube Program and the Danube River Protection Convention because their share of the land area of the catchment area is less than two percent.

The proposed Ecological Convention calls for a "holistic and integrated approach to safeguarding the overall ecological quality of the Danube Basin." It emphasizes the need to protect, preserve, and restore "the natural environment of the Danube Basin" and urges favorable consideration of regional planning and natural and cultural heritage conservation. The document is sensitive to Hungarian wariness about possible damage to its water supply, 95% of which comes from outside its borders. The threat had been symbolized by the possible damage from the Gabcikovo hydro-electric power project. At present the field has been left to the Austrian document; there does not seem to be sufficient support for a viable "ecological" agreement at this time.

21.1.2. BASIS FOR SUCCESS

The fragility of international law as an effective way to guarantee equity in the sharing of natural resources and settling disputes about them is widely in evidence. The future success of the IC depends partly on the terms of the Convention which need to be viewed with an eye to their future interpretation. The Convention recognizes many of the principles of international river basin management. It provides for the harmonizing of national policies to serve effective international environmental activities. The contracting parties are to adopt specific principles considered to be essential for the enforcement of environment policies, the polluter pays, use of best available technology, environmental impact assessment inter alia. And it affirms the need for the development and effective sharing of accurate, timely water quality data and information throughout the Basin. In the final analysis the success of the Convention and the Commission will depend on the willingness of countries to allocate scarce financial resources to its continued support and to adopt and enforce environmental standards within their own countries which are compatible with western Europe.

21.2. Other Danube Agreements - How Relevant?

21.2.1. NAVIGATION REGULATION

Like most international river basins, the Danube had had an international agreement on freedom of navigation and the maintenance of channels for transport over a long period of time. The Convention Regarding the Regime of Navigation on the Danube, signed at Belgrade on 18 August 1948, was a post-World War II attempt to re-institute navigation and other transport rules for the Danube River. Freedom of shipping on transboundary waterways had been endorsed by European powers at the Congress of Vienna in 1815 as part of the agreements which were to bring political stability to Europe for an extended period of time.

The Danube's increasing commercial importance led to adoption of a multi-lateral treaty at the Congress of Paris in 1856 and the appointment of the first European Danube Navigation Commission. Additional multilateral agreements among Austria, France, United Kingdom, Prussia, Russia, Sardinia (Italy), and the Ottoman Empire gave the Commission the power to regulate navigation for the lower course of the river and to collect dues levied on vessels leaving the river via the Black Sea.

Post World Wars I and II

The Treaty of Versailles which ended World War I extended the jurisdiction of a revitalized commission to permit freedom of navigation from the port of Ulm in Baden-Württemberg to Braila (Romania), the point where the authority of the original Commission began. At the end of World War II a conference was called to restore the Danube navigation agreement. Initially it included France, Great Britain, the United States, and the USSR, as well as the Eastern bloc countries of Bulgaria, Czechoslovakia, Yugoslavia, Romania, and Hungary. When the latter countries joined the USSR in restricting Commission membership to the seven riparian countries the Western powers withdrew from further negotiations. The five East European countries, the USSR, and the Ukrainian SSR ratified the Convention in Belgrade in 1949. West Germany refused to join; Austria became a member in 1960. The Commission's headquarters were first established in Galati, Romania and later in Budapest, where it remains. Its official languages are Russian and French.

Belgrade Declaration Jurisdiction

The Commission thus established maintains free navigation from Ulm to the Black Sea via the Sulina Channel. Each country is responsible for maintaining navigable conditions in its reach of the river and also contributes equally to a budget for the maintenance of the Commission and its services. Special river administrations were established for the Sulina channel and the Iron Gates section. Charges for the cost of maintaining navigation were levied on ships passing through these sections. The Commission has no record of a violation of its restriction on unilateral collection of transit fees from ships or other obstructions to navigation until Serb forces engaged in such practices as part of the military actions it began in 1991.

Membership Status

Membership status on the Commission has changed in the post-1989 world in a limited way. Austria, Hungary, Bulgaria and Romania have retained their status. The Slovak Republic has taken the place of Czechoslovakia and the rump Yugoslavia, consisting of Serbia and Montenegro, has replaced the federation of the past. The CIS republics of Russia and Ukraine succeeded, respectively, the USSR and the Ukrainian SSR. Croatia and Moldova have observer status, as does Germany. Ukraine is reported to object to Russian membership since Russia is not a riparian country (Linnerooth-Bayer, 1995). The Commission has held preparatory meetings regarding new memberships and other aspects of the Belgrade agreement but no changes have taken place as yet.

21.2.2. THE BUCHAREST DECLARATION

Nation states in the Danube region have a long history of bilateral and multilateral treaties. By the mid-1980s such agreements, primarily to affirm boundary locations, numbered in the hundreds (UNDP, 1984). Beginning in the 1970s more attention began to be paid to extending agreements to control transboundary pollution from air and surface-water sources. Many countries joined a number of international conventions affirming support for such goals. Danube tributary basins, including the Upper Tisza and the Drava, began to develop working agreements to curb transboundary pollution.

The agreement reached at Bucharest in 1985 to initiate monitoring at international boundary sites represented a forward step in the direction of a Danube-wide environmental agreement. It had a limited focus on water quality monitoring, however, and in general failed to use monitoring results to improve environmental policies in the riparian countries (Chapter 24).

21.2.3. THE HELSINKI AND BLACK SEA CONVENTIONS

While a rationale for its provisions could be found in a number of international agreements, the Danube Convention specifically draws attention to its relationship to two preceding international agreements: the Convention on the Protection and Use of Transboundary Watercourses and International Lakes signed in Helsinki on March 17, 1992 (Helsinki Convention) and the Convention on the Protection of the Black Sea against pollution signed on April 21, 1992. It also acknowledges the bilateral and multilateral cooperation that has taken place among the Danube countries on their own initiatives and under the Conference on Security and Cooperation in Europe (CSCE) process, and by the UN Economic Commission for Europe (UNECE) and the European Community.

The agreement has been signed by 22 countries and the European Community. The Preamble to the Convention recalls this agreement and others, including the Declaration of the Stockholm Conference on the Human Environment (1972). It references as well, in addition to these agreements the decisions made at the October-November 1989 meeting of the CSCE on the protection of the environment which was held in Sofia, Bulgaria.

Helsinki Principles

A number of principles, many of them already incorporated into EU Council Directives on the Environment, are identified to guide the overall objectives of the convention. These include the following:

the precautionary principle which supports measures to avoid environmental damage even though scientific research may not have proved a causal link between substances and potential transboundary impact;

the polluter pays principle whereby costs of pollution prevention, control and reduction are borne by the polluters;

the principle which provides that water resources are to be managed so that the needs of the present generation are met without compromising the ability of future generations to meet their needs (UNECE, 1992, p. 164-5).

The parties to the agreement are urged to:

- develop compatible relevant legal, administrative, economic, financial and technical measures to ensure pollution prevention or reduction at its source;
- use licensing of waste water discharges to protect transboundary waters from point source pollution;
- use the best available technology (BAT) to determine the limits for waste-water discharges stated in permits;
- impose stricter requirements (within their countries) when the receiving water or ecosystem requires.

Other provisions identify the need to take best available measures (BAT) to reduce nutrient inputs from industrial and municipal sources and from diffuse sources, especially agriculture; to use environmental impact assessments before approving construction and other projects which might affect natural resources, and support sustainable water resources management, while minimizing ground water and accidental pollution.

Helsinki Convention Impact

The principles and joint activities endorsed by the Helsinki Convention are for the most part included in the Danube Convention. Most of the major parts of the Danube Convention also have roots in the EU Environmental Program and the advanced environmental policies of several of the western countries. Thus, the EU Program necessitates the adoption of compatibility by basin countries applying for admission which includes all but Croatia and, of course, Bosnia and Serbia, as well as Moldova and Ukraine. The eastern bloc countries were expected to comply with EU environmental policy requirements as they changed laws and institutions to meet the criteria for admission to the Union (Part 4).

21.3. Danube Convention Provisions

The Convention on Cooperation for the Protection and Sustainable Use of the Danube River (Danube River Protection Convention), introduced in 1991 by its Austrian sponsors and negotiated during 1992 and 1993, was signed at Sofia on 29 June 1994 by representatives of all of the Danube countries participating in the Danube Program and the EC. Bosnia and Serbia are expected to be included sometime in the future. Countries with less than 2,000 sq. kms. of the total catchment area (Italy, Albania, Switzerland and Poland) are excluded. The Convention has been ratified by six of the countries; it will go into force on the 90th day after ratification by the 9th country. An interim International Commission for the Protection of the Danube (IC) has been established at the International Centre in Vienna, with its secretariat occupying offices adjacent to the Danube Environment Programme's coordination unit. When the Convention enters into force responsibility for the water-related parts of the Program will be transferred to the IC and its secretariat.

21.3.1. UNIQUE STATUS

The Danube River Basin has thus achieved a unique status among the larger international river basins of the world. It is one of the few to have (or anticipate) the commitment of all countries to achieve improved water quality through the adoption of guiding principles and a specific set of activities. Its roots in the environmental programs of the EU and, in particular, the Helsinki Convention reenforces its opportunities for success. The Convention emphasizes the specific ways in which equitable use of basin water resources can be achieved. Limited in scope to water quality issues, it nevertheless includes in its wider goals the general needs of Danube resource management.

21.3.2. GOALS AND OBJECTIVES

The Convention is a detailed but compact document which is based on the cooperative principles of international river basin management as expressed in the Helsinki and other documents and the unique needs of the Danube Basin. The references to the text of the Convention are from an unpublished document in German and English, provided by the secretariat of the interim International Commission for the Protection of the Danube River, International Centre, Vienna, Austria. "Article 13: Working languages" identifies the official languages of the IC as English and German.

The Convention's Part I states objectives, outlines its scope in terms of geographic area (see above) and planned activities. Article 3(2) identifies the planned activities and ongoing measure that are within the scope of the Convention. These include:

- the discharge of waste waters and the input of nutrients and hazardous substances from point, non-point sources, and heat discharges;
- the regulation of water construction works, run-off, and storage level control of water courses, flood control and ice-hazards abatement and

- the effect of facilities in or beside the water course;
- activities and measures for water power utilization, transfer and withdrawal;
- the operation of existing hydrotechnical constructions (e.g. reservoirs) and water power plants;
- handling of substances hazardous to water and the prevention of accidents.

The Convention is declared to apply to issues of fishery and inland navigation so far as they may be responsible for water pollution problems.

Broad Objectives

The Convention's objectives are broadly stated in Article 2 of Part I. The chief focus is on water management, which includes "the conservation, improvement and the rational use of surface waters." The countries will "cooperate on fundamental water management issues and take all appropriate legal, administrative and technical measures to improve the current environmental and water quality conditions of the Danube River...and to prevent and reduce as far as possible adverse impacts and changes occurring or likely to be caused." To this end, the urgency of water pollution abatement measures and of rational, sustainable water use, will determine priorities and the countries will strengthen, harmonize and coordinate measures at the domestic and international level for sustainable development and environmental protection of the River throughout the basin.

Following the lead of the Helsinki Convention, the Danube agreement emphasizes the need to adopt the polluter pays principle and the precautionary principle as basic to all provisions. Signers have agreed to support 1) conservation, improvement and the rational use of surface and ground water in the catchment area; 2) to control the hazards originating from accidents involving substances hazardous to water, floods and ice-hazards, and 3) to contribute to reducing the pollution loads of the Black Sea from sources in the catchment area (Art 2.1). They will also work to "strengthen, harmonize and coordinate measures taken and planned to be taken at the domestic and international level throughout the Danube Basin aiming at sustainable development and environmental protection of the Danube River (Art 2.3)."

Achieving Goals: "Multiple cooperation"

Part II of the Convention sets forth the action necessary to achieve these objectives. The parties to the agreement are expected to take legal, administrative, and technical measures to prevent, control and reduce transboundary impacts (Article 5). They are to record conditions of natural water resources within the catchment area, applying agreed-upon quantity and quality parameters; to adopt legal provisions which set time limits to be met by waste water discharges and for handling substances hazardous to water and for reducing inputs of nutrients or hazardous substances from nonpoint sources. The IC will suggest how the implementation of regulations should be harmonized among the countries.

Article 6 prescribes specific water resources protection measures, identifying the groundwater resources subject to long term protection and protection; prevention of pollution of groundwater caused by nitrates, inter alia; reduction of risks from accidental pollution; taking into account possible influences on the water quality resulting from planned activities (Article 3 specifies construction works, among other activities); and the evaluation of the importance of different biotope elements for riverine ecological conditions.

21.3.2. EMISSION LIMITS

The two western Danube countries, Austria and Germany, have adopted and enforce emission limits for the discharge of waste into surface water. The IC will propose such limits for adoption by the countries, "applicable to industrial sectors or industries in terms of pollution loads and concentrations and based in the best possible way on low- and non-waste technologies at source." For hazardous substances emission limits will be based on the BAT for abatement at the source and/or waste water purification. For municipal waste water, emission limits are to be based on the biological or an equivalent level of treatment. As a counterpart, similar provisions are to be developed for non-point source pollution, particularly from agricultural sources

A detailed list of industrial sectors and industries and of hazardous substances to be reduced or prevented is contained in Annex II of the Convention. The Commission is expected to update these lists. General guidance for the definition of water quality objectives and criteria to prevent, control and reduce transboundary impacts is also contained in Annex III.

Article 7 specifies the "necessary preconditions and implementation" that the parties should ensure in order to reach "an efficient limitation of the emissions" in their respective jurisdictions." There are a number of detailed provisions covering regulations for emission limitations, permit discharge systems, the imposition of more stringent provisions under some circumstances, environmental impact assessment, among others.

Emission inventories are to be undertaken periodically at the relevant point and non-point sources of pollution within the catchment area to register the measures taken with respect to the discharges and their "actual efficiency (Article 8)." This information will support lists of further prevention and abatement measures to reach the overall objectives of the Convention. The inventory and list of measures to be taken will "form the basis for joint action programs, taking into account priorities based on urgency and efficiency." Periodic progress reviews will monitor the joint programs

22.3.3. MONITORING, DATA AND INFORMATION SHARING

The Convention's first major premise, that resource protection will not be achieved without the basin-wide enforcement of compatible water quality standards and emission limit values, has supporting provisions for the establishment of a program to monitor the results of water quality protection throughout the basin. Monitoring must be based on a harmonized methodology which will result in the establishment of "domestic water balances, as well as the general water balance of the Danube River Basin." These data should make periodic assessment of quality conditions possible as well as "the progress made by their measures taken aiming at the prevention, control and reduction of transboundary impacts." Information about monitoring results will be made available to the public in appropriate publications.

Reporting Obligations

Article 10 specifies each country's reporting obligations to the Commission. So that the Commission may carry out its tasks it needs information on bilateral and multilateral agreements, national laws and ordinances and other stipulations "regulating the protection and water management of the Danube River and of waters within its catchment area or being relevant for questions concerning them." The countries will offer suggestions for the implementation of Commission decisions, including time frames and costs. The countries will also report planned activities which might cause transboundary impacts. If there are problems with such impacts consultations with the Commission will take place (Article 11).

Exchange of information and "reasonably available" data is provided for in Article 12. In particular, "to harmonize emission limits, the Contracting Parities shall undertake the exchange of data on their regulations." Other items include "emission and monitoring data," and "accidents involving substances hazardous to waste," BATs, etc. Confidentiality of information is to be respected (Article 12 (4) and (5)). "The provisions of this Convention shall not affect the rights or the obligations of Contracting Parties in accordance with their domestic laws, regulations,administrative provisions or accepted legal practices and applicable international regulations to protect information related to personal data, intellectual property including industrial and commercial secrecy, or national security."

Article 13 protects "industrial and commercial secrets or other confidential pieces of information" which may be made available. Such information cannot be used for any other purpose than those stipulated in the Convention, and cannot be published or made available to third parties.

Public Information

Contracting parties are to ensure that they provide information to the public on the quality of the riverine environment of the basin "in response to any reasonable request, without that person having to prove an interest, as soon as possible." Countries may refuse such information for reasons of public security, or confidentiality of personal information.

21.3.4. RESEARCH PROJECTS

The Convention also provides for complementary or joint research and development programmes and the sharing of the results (Article 15). Article 16 provides for communication, warning and alarm systems, and emergency plans. Along the same lines, mutual assistance procedures are encouraged and the Commission will elaborate them (Article 17).

21.4. The IC, Prerogatives, Structure

International river basin conventions are inevitably concerned with one key issue: the nature of the authority granted to the administrative agency (Commission) that will implement its provisions. "For implementing the obligations of the contracting parties pursuant to Articles 1 to 18 the International Commission elaborates proposals and recommendations addressed to the Contracting parties (Article 18)." In addition to decisions explicitly entrusted to the IC it can be granted additional responsibilities by the contracting parties within the scope of the Convention's objectives.

Each party nominates five representatives, a head of delegation, a deputy, and experts. The IC chair is rotated annually among the parties in alphabetical order (as determined by the language of the country) with one member of the chair's delegation serving as President of the Commission. It is to meet once a year, although meetings can be convened by the President at the request of at least three delegations. Each delegation has one vote. "The European Community, within the areas of its competence, is entitled to a number of votes equal to the number of its Member States which are Contracting Parties to this Convention." When member states are voting, the EU will not vote on their behalf (Article 4). Decisions are ordinarily to be adopted by consensus but can be made by 4/5ths of the contracting parties.

During the negotiation process for the Danube Convention it was decided that work of the IC should begin on an interim basis. During this period and until the Convention has entered into force the interim secretariat is legally based on an Austrian law which was enacted during the summer of 1995. Rules of procedure were adopted and a transition period will be used to modify and activate complementary work groups established by the Danube Program.

CHAPTER 22

Database And Information Systems

22.1. Significant Needs

The Danube Program began its ambitious assortment of activities in 1992 without accurate, compatible data about the environmental status of the basin. Basin countries lacked an electronic system which would inform them of accidental spills and other emergencies despite the lessons learned from the Sandoz disaster on the Rhine (Chapter 4). Data made available by each country on the levels of toxic pollutants in tributaries and the main stream as well as in water supply systems lacked a common format and in many countries, significant substance. Electronic access was not available to such data or to information about other aspects of environmental policies. Monitoring begun under the Bucharest Declaration in 1988 measured the levels each month of some thirty parameters at eleven transboundary sites. Sampling and laboratory methods were not compatible and the data were shared on a yearly basis, long after the measurements taken during the final month of the previous year. The results were not used for policy decision purposes.

22.2. Designing Danube Systems: Complex but Doable

Work began in 1993 to provide an accident warning system and to update and make compatible monitoring, laboratory and information management systems throughout the basin. Formidable challenges faced sub-group experts as they began work. Lack of a common language was one. Incompatible equipment and expertise were other handicaps as was the gap created in networks by lack of participation by two major Danube countries, Bosnia and one of the two countries left in the federation, Serbia.

22.2.1. ACCIDENT EMERGENCY WARNING SYSTEM

The development and dissemination of an accident emergency warning system, an item on the short-term, high priority action agenda of the Danube Program, was the responsibility of the AEWS sub-group of the Task Force. Dr. Gyorgy Pinter, Deputy Director of VITUKI, the Hungarian water research institute, chaired the sub-group. It worked separately from the new sub-group on Monitoring, Laboratory and Information Management (MLIM, see below), and produced a separate system. The AEWS sub-group had technical support from Delft Hydraulics. Lack of compatible international or national data on accidents and spills and their sources was an initial problem for the group. In any case all studies showed that there was an urgent need for a warning system.

By the end of 1995 the eleven participating countries had established a network of National Principal International Alert Centers (PIACs) to coordinate emergency warning at the international level. Satellite communications equipment, information processing systems, and hazardous substances databases are in place; training in their use was expected to begin in 1997 and the official opening of the system to take place shortly thereafter. It was recognized that the different languages used in the basin would create problems and so software serving the operation of the PIACs has been translated into local languages.

A consortium led by VITUKI, the Hungarian Water Resources Research Institute, is developing a "first version of the Danube Basin Alarm Model." The model will be able to predict the arrival time and concentration of accidental pollution in the River and its tributaries. It will help the PIAC staffs to decide when to take action, for example, shutting down drinking water intakes. Based upon the Rhine River Alarm System, the Danube model was demonstrated in workshops in 1995 in each of the eleven participating Danube countries. Government authorities were informed about the structure and operations of the international system and provided detailed information about the respective roles of institutions and experts.

22.4. Monitoring, Laboratory and Information Management

The work plan adopted in 1992 by the Danube Program Task Force identified three activities for improved monitoring and data management:

- the development of an information system with national standards and relevant legislative texts with long term action to be undertaken within the context of the Convention;
- improvement in environmental monitoring, sampling, selection of parameters, and strengthening of laboratory methods;
- the development of a strategic and comprehensive approach to data collection, processing, and exchange for the basin which would include decision systems, and
- computerized infrastructure for efficient data input, use, and exchange, with the results to be made available to the public.

22.4.1. SUB-GROUP FORMED

A sub-group was formed by the PCU of the Danube Program to carry out the MLIM mandate. Initially, a consortium, know as WTV, was appointed by the European Commission to evaluate the existing capabilities of the basin with respect to the goals of the sub-group on monitoring, laboratory analysis and information management. The recommendations of the consortium were that a Trans-National Monitoring Network (TNMN) be established so that the chemical and biological quality of the Danube might be uniformly assessed. National Reference Laboratories were to be established in each basin country, and an information management system set up for the TNMN to support the exchange of data, statistical trend analysis and interfaces developed with national information systems in each country.

The project was to strengthen existing environmental monitoring networks, improve the comparability of sampling techniques and laboratory analyses. It was to build on and, where necessary, develop and implement compatible information management systems. The ultimate goal: fulfill the management needs at the local and national level and internationally (Lack, 1996).

Three working groups of specialists were assigned to advance these goals: one concerned with monitoring, one with laboratory management, and one with information management. The groups worked intensively to resolve a complex set of problems, with frequent workshop sessions over a three-year period.

22.4.2. MONITORING

The monitoring work group assessed current monitoring practices in each Danube country to determine their ability to meet long term and emerging water quality monitoring needs. Data were needed which would permit evaluation of water quality levels as they compared with accepted international standards. Trend data, noticeably lacking in the pre-transition Danube River and its tributaries, were needed to evaluate the effectiveness of environmental management policies over time.

The TNMN used the monitoring stations of the Bucharest Declaration monitoring program and increased the total number of stations to 49. Following an evaluation of their effectiveness, it is expected that the number will be increased to 79.

A number of decisions were reached at a May 1996 meeting of the sub-group. A list of the determinands (parameters) to be monitored and the frequency of sampling was drawn up. All countries (excluding Bosnia and Serbia) were said to be ready to comply with the first phase of the monitoring activity if not by 1996, by 1997. Unresolved issues included the need to improve the collection and integration of flow and water quality data and to emphasize the calculation of loads of contaminants.

22.4.3. LABORATORY MANAGEMENT

This group worked on the establishing compatible analytical systems in each country where disparate methods had previously been in use. Sampling and field monitoring equipment for certain determinands is being acquired and distributed. The establishment of a National Reference Laboratory has been recommended as a central coordinating unit for Network sampling and analysis methodology. In addition to equipment purchases there is a need for equipment such as water de-ionisers and for consumables such as reagents and gases. The laboratories need to have improved analytical quality control and to be able to carry out performance testing. International checking of sampling and laboratory management at regular intervals is also anticipated.

Processing of water quality data obviously required more than the collection of laboratory analyses; data need to be transformed into information that will be of use in a variety of circumstances. This is to be accomplished by a Data Exchange File Format (DEFF) which will interface with each national information system in parallel with the national reference laboratories. The software-based protocol has five related tables of information: monitoring points, analyses, remarks, methods, and determinands. It is designed to facilitate the transfer of water data internationally according to the experience and capabilities of Danube countries. For the present, each country will hold the data for ready access, making use of it as it sees fit. "There is a need for an international working group to collect, collate and process the data from the TNMN working in a similar way to the Rhine Secretariat of the International Commission for the Protection of the Rhine (Lack, 1996)." Major training and research programs in water quality monitoring and data and information sharing will continue.

22.4.4. INFORMATION MANAGEMENT

Agree-on Objectives

The need to provide access to data and information about the basin to a wide range of users was recognized in 1991 when the Danube Environment Program was launched. The founding meeting in Sofia in September of that year held discussions on the subject and had demonstrations of systems then available. While the GEF reported in the Danube Program's 1994 Annual Report that a system that would fulfill the goals of the Danube Program had been designed and initiated with the support of UNESCO, this system apparently is no longer in use. In any case, it did not appear to have fulfilled the wide-range goals defined in early Task Force meetings.

The 1995 Annual Report of the Danube Program described a new system funded by the GEF called DANIS (Danube Information System) is described as having three objectives:

1. to provide access to all existing primary information systems and databases developed for the needs in the Danube region;
2. to act as an information source assisting the PCU and the Country Programme Coordinators in their daily tasks;
3. to provide information to the public, NGOs and other interested parties in an efficient way.

Achieving Objectives

By 1996 none of these objectives had been achieved. Access has not been provided through this information system to any of the several existing information systems and databases. As an information source it could well be assisting the PCU and the CPCs but it is not clear how this is being accomplished. The titles of its databases are vague and in some cases do not accurately describe their contents. For example, "legislation," lists documents which are actually treaties and have nothing to do with legislation. The documents themselves cannot be accessed. It was clear in 1996 that DANIS needed help if the objectives were to be achieved.

As an example of databases that are available from other sources, there is the EBRD and the Phare program database of standards for eight media and subjects in 22 countries, and those in CEC directives and WHO publications. The latter contains only recommended criteria for international use since no entity has the authority to develop enforceable international standards for environmental media. This database, which is available from the Central European University in Budapest, was demonstrated to the PCU following its completion in 1994.

22.5. NATO Workshop

A re-evaluation of the status of data and information sharing in the basin was conducted at a NATO Advanced Research Workshop in Budapest in May 1996. Workshop participants included the former team leader of the Danube Program PCU, David Rodda; Tim Lack, who has led the MLIM sub-group; representatives of ministries of environment in each Danube country with the exception of Moldova whose ministry submitted a report to the workshop; a Rhine Basin Commission member, and four NATO country specialists, as well as experts on the current status of data-sharing among Danube countries. They reviewed the activities of each work-group and recommended an action strategy for the optimal sharing of water quality data among Danube Basin countries (Murphy, 1996). Dr. Lack facilitated the development of recommendations for a coordinated approach to data and information sharing in the Basin.

After considering the eco-system needs of the basin these specialists contributed to discussions of current environmental monitoring in other international river basins, water quality monitoring and data-sharing systems in four NATO countries and in the basin countries in transition. They reviewed the work of the five-year Danube Programme and the needs of the soon-to-be permanent secretariat of the IC. They watched demonstrations of information and data systems available on diskette and the Internet.

The group reached consensus on a number of recommendations. It was agreed that monitoring data from the TNMN network should be matched with other sources of data and information in order to provide a single point of enquiry that will allow all user requirements to be met. An Internet application based on the world wide web (www) was recommended which would be technically simple, relatively low cost, and would allow dissemination at a global level. Key recommendations were followed up in a NATO Linkage Project which brought together the University of Ljubljana, Colorado State University in the US, and Central European University in Budapest in the first six months of 1997. It was expected to result in electronic products which will be put in place for use by the secretariat of the IC.

CHAPTER 23

Civil Society and the Environment

23.1. Basic to Democratic Reform

Support for democratic reform required fundamental changes in political decision processes in each of the former socialist countries in the Danube Basin (Chapter 10). Special help for the improvement of environmental policies came from a number of sources. Incentives to adjust political and economic policies re-enforced adoption of western environmental norms and methods to enforce them. The Danube Program emphasized "institutional strengthening," defined as support for governments as they struggled to embrace such new institutions as free access to environmental data and information and the protection of groups which represented interests independent of the government.

23.2. Help to Countries

Beginning in 1995, the GEF provided funding to the Danube Program Coordinators in each country who were responsible for carrying out Program activities (EPDRB, 1995, p. 7). Coordinators were usually individual specialists, housed within the ministries of environment, who helped to make and carry out decisions of the DEP Task Force and PCU. Funding provided by the GEF supported additional staff for activity coordination. This move responded to a common complaint from the ministries in the former eastern bloc Danube countries, i.e. that they were expected to provide the time of their experts and facilities without compensation while international agencies funded western consultants for their part of the work.

23.2.1. SUPPORT FOR MINISTRIES

Help was provided for the coordination of environmental policy among ministries, particularly those in charge of industrial and agricultural development and public health. Seminars in environmentally sustainable industrial policy brought ministerial representatives from Bulgaria, Moldova, Romania, and Ukraine together who concluded that a lack of investment incentives, the current macro-economic situation, and uncertainties in the regulatory framework were retarding industrial restructuring and leading to sub-optimal environmental benefits. Improvement would come when ministries in each country shared work programs, developed a better understanding of licensing and permitting of industrial waste emissions, and training in the adoption of "win-win" reforms by government and industry which would pay for themselves. A second workshop was held for the upstream countries, the Czech Republic, Croatia, Hungary, Slovakia, and Slovenia, which focussed on the licensing of industrial facilities. Workshops on sustainable agricultural policies were also provided.

23.2.2. LOCAL GOVERNMENTS

Local governments received training in 1994 and 1995 with emphasis on water sector issues, harmonization with EU environmental regulations and standards, and waste water treatment, and other water services activities. The PCU also promoted exchanges between German waste water treatment plant managers and those in the basin. A review of river basin management practices in the Danube countries linked the basin planning content of the National Action Plans which were to help implement the Strategic Action Plan with recommended improvements in institutional structures and financial arrangements for integrated water management.

23.3. Aid to Public Participation, 1994-1996

Public participation in government decision making, a function of the "Civil Society," followed a common pattern in east European countries in the early years of the transition. In most of the socialist countries non-governmental environmental groups (NGOs) developed as appendages to communist governments, funded and controlled by the communist party. Many advocated protection for natural resources but did not act independently and did not constitute a civil society component as it was understood in democratic countries. Response to constituencies and impact on government policies were minimal.

23.3.1. SHIFT IN NGO GOALS

With the end of the Cold War many NGOs had recruited members from severely polluted areas and become politically active. For a time it was difficult to distinguish them from political parties or units under the control of political parties. Ecoglasnost in Bulgaria and TER (the Ecologist Youth of Romania) were, for example, active in the immediate post-1989 period, and each, allied with other environmental non-governmental organizations, helped to elect some pro-environment candidates. The return of formerly Communist leaders in 1992 identified such organizations as "the opposition" and did not give a high priority or ministerial positions to their members.

Regional Environmental Center, Budapest

As these groups shifted from dependence on government subsidies (such help continued at a reduced level in most of the formerly socialist countries) to greater independence, help came from the west in various forms. A major source of funding was provided by the Regional Environmental Center for Central and Eastern Europe (REC) which was opened by US President George Bush in September 1990. Funding of $5 million was committed to the center for operational expenses and grants. The source was USAID but funds were allocated through the US Environmental Protection Agency. Legal complications required that REC be organized and operated under the laws of Hungary; it did not become an international entity until 1995. EU Phare committed an additional $3 million and European governments and the Japanese government made further contributions.

Other Sources

Not all support for public participation traveled through the Center. Immediately preceding and following the post-1989 changes in governments help came from foundations in the west, including the US. The GEF made its first contributions for activity support in 1995 by instituting a Danube Grants Program.

Initially, international agencies received money for meetings and workshops which would help the nascent civil society in the Danube countries. REC was criticized by NGOs in countries other than Hungary for its first-year grant policies which allocated over 60% of funding to Hungarian groups. Funding for Romanian groups concentrated on Hungarian organizations in Transylvania. Help with the coordination of projects through a revised management plan came from the EU Phare program. Program managers were brought in from Bulgaria, Romania, the Czech Republic, and Poland and after a time coordinating offices were established in each of the countries eligible for assistance.

23.3.2. NGO DANUBE FORUM

Improved coordination was the object of a REC-funded NGO Danube Forum in 1993. Its concept had been suggested at the first meeting of the Danube Program Task Force in 1992 by Delcho Vitchev, a leader of the Bulgarian NGO Ecoglasnost who had helped to overturn the communist government in Sofia in 1989. Mr. Vitchev suggested that the Task Force needed to sponsor a "Forum" which would guarantee wide participation in the region and not favor any one group or any one country. The Forum would remain essentially unstructured but would develop plans to strengthen NGOs to help with the transition from government control to independence and true representation of constituents.

Workshops in each Danube country resulted in a database of detailed information about NGOs and a consensus on future needs for public participation in basin environmental policies. The workshops recommended information centers in each country that would serve the maximum number of NGOs at the lowest cost. Coordination of assistance would be helped and favoritism toward any one group would be avoided by rotating the direction of the centers among organizations. There was no immediate response to the recommendation by the REC or by the Danube Program coordinators.

23.3.3. TOP-DOWN OR GRASS-ROOTS GRANTSMANSHIP?

The GEF funded a meeting in Sinaia in February 1994 and in Mikulov in the Czech Republic in October of that year "to establish a broad, basin-wide network addressing environmental problems and needs, and to discuss the principles and means of public participation and awareness raising (Paranici and Petrescu, 1994, p. 1)." The network was to be called the Danube Environmental Forum. Instead of information centers, it was decided that chairpersons of three "sub-regional focal points" Czech Republic, Slovenia, and Romania, should become members of and represent NGOs on the DEP Task Force in May 1994.

These representatives have received no funding; an informal survey of major NGOs in each country suggests that the representatives do not contact their clientele (presumably NGOs throughout the basin) and do not raise environmental issues with ("lobby") national or international policy makers. The author did not receive replies to repeated enquiries directed to the three NGO task force members. According to Ms. Annie Roncerel, then the GEF coordinator for the Danube Program, this was probably because the NGO representatives had never received funding for their activities (Roncerel, 1996). An informal survey of NGOs affirmed the need for information centers in each country.

23.4. GEF Grants Program

GEF disbursed 108 awards totaling $216,855 for a Danube River Grants Program in 1995 (EPDRB, 1995 Annual Report, p. 11). The Report indicates that financial support has been provided for two NGO information centers, one for the upper river basin (Union of the Morava River, $3,000) and one for the lower (Group for Speleologtical and Underwater Exploration in Bucharest). Requirements for the grants were few in number. NGOs had to have been organized for at least a year and to subscribe to grant program goals. The NGO Danube Program task force representatives were not involved in the allocation.

The World Wildlife Fund, which also is represented on the Task Force, has a "Green Danube Program" which addresses eco-system problems caused by channelization of the river, draining of wetlands, building of dams, cutting-down of forests, and pollution. It has a mix of field projects, public education activities to help achieve long-term goals of ecological restoration and conservation. Presently five projects demonstrate model forest and meadow management, the restoration of drained wetlands, protection of sensitive habitats, and ensure the survival of endangered and threatened animal and plant species. The program is funded at 220,000 Swiss francs ($165,000).

23.5. Future Plans

GEF provided funding for 1996-1997 includes $3.9 million for the development of Danube Basin projects and $1.79 million for a strategic action plan for the Black Sea countries (Bulgaria, Romania, Russian Federation, Georgia, Turkey and Ukraine).

Funding goals for public awareness are broadly presented: "Participation of all sectors of society is an essential requirement for the developments of sustainable policies in the Danube Basin." There will be education projects, support for transparent and participatory decision making procedures, and for access to administrative and judicial procedures. Participation at all levels requires open exchange of data and information. A total of $590,000 for these purposes has been allocated. The GEF coordinator, Ms. Annie Roncerel, has promised an objective evaluation of the Danube Grants Program and other aid provided for public participation in Phase I of the Danube Program before specific programs are planned for the funding for 1996-1997 (Roncerel, 1996).

An Applied Research program funded by the EU Phare program indirectly strengthens the civil society. The goal of democratic decision making is served if the researchers formerly responsible to one-party governments can become independent. They support scientifically sound contributions for the establishment of goals for public policy and can provide the public with objective information.

Thirteen projects have been funded under the Applied Research program. They were designed to help to evaluate risks, establish inventories of environmental protection needs, accomplish baseline studies of biodiversity, develop monitoring methods linking interaction between quality of sediments and surface water, develop improved environmental quality standards, provide a deeper understanding of sediment nutrients, and help to create computer models for use in predicting pollution dispersion especially after spills. Of 38 institutions identified for possible funding 29 are from Danube countries and nine from European Union. A total of MECU 4.273 have been allocated for the program.

CHAPTER 24

Investing Strategically in the Basin: 1995-2005

24.1. Strategic Action Plan Adopted

A plan for the sustainable use and development of an international river basin must be adopted, and implemented, if conflicts between countries are to be reduced and equitable sharing of resources to take place. The first Danube Program Task Force meeting in February 1992 proposed a "Strategic Action Plan" (Action Plan, Plan) which was less a use and development plan and more suggestions for waste treatment projects. It emphasized the need for an inventory of the most heavily polluted areas and identified steps to restore and protect them in the future.

The lack of accurate data on the status of the environment made precise identification of investment needs difficult. Available data and information came from national and international sources, including World Bank reviews, reports to the UN Conference on Environment and Development in Brazil in 1992, the national reviews requested for the Program, and statistics developed within the countries. The plan was put together in 1993 and endorsed by all participating basin countries in May 1994 at a meeting in Bucharest.

Early discussions with each riparian country assigned the highest priority for clean-up projects to fifteen tributary basins. Initial conclusions cited needs which had already been anticipated: principal environmental problems resulted from poorly treated or untreated municipal and industrial waste, frequently mixed together in overloaded municipal waste systems. The next step was to establish, in consultation with the countries, "investment pipelines" for funding. It was difficult then and remains difficult now to establish the exact cost of the items included in the Plan. Sums in the billions were anticipated (EPDRB, 1993). Country and donor representatives agreed, in any case, that securing funds would be extremely difficult.

24.2 Plan Objectives

The Plan set broad objectives for short term (by 1997) and medium term (by 2005) achievements. Three long term goals (no time set) presumed that all of the intermediate objectives had been achieved and permanent sustainable development attained. Immediate objectives guided initial investments in environmental management, setting a working basis for strategic and integrated management of the Basin environment. While it refers to other ecological problems, the Plan is water quality and quantity oriented.

24.2.1. MAJOR GOALS

Its major goals are to:

1. reduce the negative impacts of pollution in the basin and on riverine ecosystems and the Black Sea;
2. maintain and improve the availability and quality of water in the basin;
3. control hazards from accidental spills, and
4. develop regional water management cooperation (EPDRB, 1994, p. v).

The Plan has an ambitious secondary goal: to support "the transition from central management to a decentralized and balanced strategy of market-based incentives."

After its adoption it was agreed that the countries would review the Plan's progress in three years. In the meantime its implementation would be coordinated with the Danube Program PCU and the IC.

24.2.2. WATER QUALITY AND QUANTITY ISSUES

The Plan addresses four major water pollution issues:

the high load of nutrients and eutrophication in the main river and its tributaries;
presence of hazardous substances, including oils;
microbiological contamination, and
substances causing heterotrophic growth and oxygen depletion.

Increasing competition for useable water was a fifth major concern. Each of these factors limited water uses of all types: drinking water, fisheries, industries, irrigation for agriculture, and recreation. (See individual profiles above in Part Four under Environmental Issues.)

The Plan's objectives respond to subsets of these problems in two ways. Priority targets are described in abstract terms. More specifically, "hot spots" are identified in a table which gives their geographical sites, the type of facility, and suggests the work needed and, in some cases, the probable cost of upgrading or replacing facilities. There are 179 such sites listed in nine countries.

24.3. Targets and Schedules

24.3.1. SHORT-TERM

An initial short term target asked each country for a National Action Plan (NAP) which identifies how it will fulfill its responsibilities under the master plan. A time and cost schedule is requested. Overall Plan requirements are ambitious and general. For example, following the preparation of the NAP each country is to prepare integrated tributary river basin plans and revised water allocations and water use permit systems. The targets for 1997 are itemized below not because they are in any way realistic but because they do capture what should, under ideal political, economic, and cultural conditions take place.

Targets for 1997

Targets include:

- completion of wetland inventory conservation and management programs; adoption of consistent water quality objectives and criteria for all Danube
- tributaries and the main river;
- completion of a regulatory and permitting reform program for water use and industrial enterprises;
- adoption of emission limits for fertilizer plants and of emission limits based on best available techniques (BAT) for new industrial sources and livestock enterprises;
- completion of a comprehensive system of information on the state of the riverine environment;
- national assessment of critical loads and load reduction targets for the highest priority river reaches;
- evaluation of the critical load of nutrients from the Danube into the Black Sea;
- completion of effective and comprehensive monitoring, warning and laboratory systems, including systems for the exchange of information, and
- development of technical and management capacity building programs for all actors and sectors.

24.3.2. MEDIUM TERM TARGETS

After reaching these targets the countries should achieve medium term goals by 2005:

- completion of a pollutant emission inventory; adoption and implementation of hazardous substance control legislation including control of transport pollution;
- introduction of regulations for fertilizer storage, handling, and application;
- preparation of waste water and sewerage investment priorities for cities, rural towns and villages;
- completion of the rehabilitation and modernization of existing municipal waste water treatment plants;
- investment in highest priority sewerage and municipal waste water treatment capacity expansions;
- completion of projects on the conservation and restoration of priority wetlands;
- introduction of environmentally sound agriculture policy reforms;
- demonstration of Best Environmental Practice (BEP) for the use of fertilizers, pesticides, and other agrochemicals in agriculture;
- completion and application of pilot and demonstration projects for manure handling, storage, and disposal;
- introduction of phosphate-free detergents and ban on phosphate-containing detergents, and
- phased application of emission limits and incorporation of emission limits into

24.3.3. LONG TERM TARGETS:

Long term targets were few but ambitious:

- completion of the construction of municipal and industrial waste water treatment plants;
- change to sustainable agricultural practices, and,
- restoration of natural purification capacity of the Danube River and its tributaries.

24.3.4. MEETING TARGETS: A METHODOLOGY

Building on its identification of basin water quality and quantity problems, the Plan proposes sets of "actions," which will also be short and long term and that need to be undertaken by the countries themselves. Danube counties are urged to do the following:

- prepare a list of costed, immediate priority actions addressing major health, ecological and economic risks and identify "who" should take the priority actions, how the financing will be obtained and a time scale;
- identify a mechanism and the necessary resources to support the production of the NAPs which each is urged to complete immediately;
- assess the public health and ecosystem risk of not meeting the short and medium term targets in the Strategic Action Plan.

"Actors" are identified who must develop and implement solutions. The list includes government authorities, municipal water companies (presumably also under the regulation of the government), "industrial enterprises" "the general public and NGOs." The list of responsibilities is long, generally unrealistic and lacks an understanding of decision processes. Many are "short term" to be completed in 1995-1997.

24.3.5. NATIONAL ACTION PLANS: COUNTRY RESPONSIBILITIES

Four primary goals are established for the NAPs. Their preparation is closely related to the National Environmental Action Plans (NEAPs) requested for the completion of the Environmental Action Program for Central and Eastern Europe adopted by the Lucerne Ministerial Conference in 1993 (Regional Environmental Center, 1995). The NEAP requests the outline of an environmental protection strategy which is consistent with a market-based economy and includes planning for the use of all natural resources.

The Plan suggests that the PCU of the Danube Program and the secretariat of the IC should play coordinating roles, arranging for expert assistance, meetings and workshops, inter alia. The international community would help in funding priority actions but long-term financing would have to be provided by the countries themselves. And the decision processes within each country for each level of government would have to satisfy a large number of demands for vast improvements in the management of water quantity and quality.

The Plan, in effect, set goals for government decisions (Figure 3, chapter 10) and offered help with the process of reaching them through legislative and administrative action by national governments. Parliaments and cabinets in the Danube countries, except for Austria and Germany, were to enact and administer legislation to support policy reform while they also allocated resources to fund clean-up projects. The work cut out for them by the Plan was formidable in its dimensions, complexity, and cost at a time when competition for the resources of all the Danube countries was severe.

24.4. The Plan's Implementation Program

As a way of recognizing the overly broad provisions of the Strategic Action Plan, a Danube Strategic Action Plan Implementation Program for 1996-1999 was introduced in 1995 to suggest more specific activities for the Danube countries. These were included in proposals for funding submitted to the EU Phare/Tacis programs and to the GEF in July 1996. Tributary basins specifically identified for assistance included the Sava, Mura/Drava, Morava, Vah, Tisza, Prut/Siret, Jiu and Yantra.

The SIP was to "make a positive and concrete contribution to the rehabilitation of the Danube environment, up to and including the receiving waters of the Black Sea." It was to be "driven by the Danube countries themselves," which were to offer proposals for the implementation plan (EPDRB, 1996 p.3). It was endorsed by the ministers and senior officials of Danube countries at the joint Task Force meeting in July 1996.

24.4.1. INTERNATIONAL DONOR SUPPORT

Donors were urged to give their full support, by way of grants, to achieve six goals. These include:

- Concrete progress in reduction of pollution from urban and industrial waste waters through preparation of several waste water investment projects;
- Basin-side dissemination of investment methodology and methods of financing capital projects;
- Working with farmers to achieve improved and sustainable farming practices and methods of manure management from intensive animal farms leading to basin-wide pollution reduction from agriculture;
- Immediate benefits to demonstration wetlands projects and future benefits to other wetlands;
- International cooperation in sustainable management of selected transboundary rivers and improved structure and capacity of river basin management institutions;
- Working with communities to reduce pollution and improve water services in rural areas by raising awareness and generating investment projects for financing;
- Activation of community based constituencies and mobilization of the NCO community for the work of the IC of the DRPC.

The proposals focus on eight of the main Danube tributaries, selected on the basis of their transboundary importance and their representative characteristics to ensure that SIP results will be replicable throughout the basin, prepared in the light of GEF and Phare funding criteria and address the priorities and objectives of the Plan.

Table 24.1 "Hot Spot" Cost Estimates in Danube Basin ($millions)

	WWTP Mun. etc	Effect Dams	Ind, Ag Mining	Costed, WWTP	Costed, Other	TOTAL
Bulgaria						
Number	20	1	8	2	NA	29
Cost				67.9	3.1	71.0
Croatia						
Number	18	NA	4	2		22
Cost	NA			1.7	2.3	4.0
CR						
Number	5	NA	5	4	4	10
Cost	NA		NA	60.9	5.2	66.1
Hungary						
Number	20	NA	9	19	1	29
Cost	NA		NA	821.0	129.5	950.5
Romania						
Number	15	NA	28	NA	NA	43
Cost	NA		NA			NA
SR						
Number	9	NA	13	NA	NA	22
Cost						NA
Slovenia						
Number	11	NA	2			13
Cost	43.0		NA	43.0	0.5	NA
TOTAL NO.	87	1	69	27	6	
TOTAL COST				994.5	140.6	1,135.0

CHAPTER 25

Transition to the Commission

25.1 A Planned Transfer

With the final ratification of the Danube Convention expected by 1997, the Danube Program Task Force prepared to hand over its responsibilities to the IC. Two major problems had to be resolved. An appropriate transfer protocol had to be developed and implemented. And funding sources for the operation of the new Commission had to be identified.

A preliminary meeting of Country Program Coordinators in November 1995 discussed both subjects. Ilya Natchkov, Bulgarian Ministry of Environment, who had been participating in the Task Force since its first meeting in 1992, chaired an ad hoc working group on the transfer of responsibilities and tasks. His report was reviewed during the two-day meeting and then formally acted on at a joint meeting of the Program Task Force and the interim IC in July 1996 (EPDRB, 1996a). A Program Management Task Force (PMTF) was established on an interim basis to carry on the responsibilities of both the Task Force and the Commission pending the entry of the Convention into force.

The PMTF included the contracting parties of the Convention, donors, and the international financing institutions. Participating nongovernmental organizations were to be identified later. A Program Coordinator, eventually to be appointed by IC members, will carry out an action program drafted by the PMTF. Program managers will be appointed, as at present, by specific donors and/or international funders. The IC chair will rotate among the countries of the contracting parties (Danube countries), in alphabetic order (language of the country).

The interim IC was represented by a temporary secretariat at the Vienna headquarters. Its President was Wilhelm Kittinger of the Austrian Federal Ministry for Agriculture and Forestry. The secretariat included a technical director, Prof. Hellmut Fleckseder, of the Federal Ministry and an executive secretary, Gerhard Velcovsky, of the Austrian foreign ministry. The secretariat has an office adjacent to the present PCU at the Vienna International Centre, Vienna, Austria.

25.2. Funding Background

Initially, the Danube Program was funded from 1992 to 1995. Continued program funding, for a second phase to extend from 1996 to 1997 or 1998 has been committed or promised by the two primary contributors to the Program, the GEF and the Phare program. Funding for the permanent IC and its secretariat is still uncertain, as is support for the substantial capital investments needed for waste treatment plants and modernization of plants and other facilities to improve their treatment capacities.

In addition to grant money provided to the Program (see Table 25.1) the EU Phare Program gave grants directly to east European countries for improvements in environmental programs. Assistance, including grants and loans, made available by various sources including the EBRD and the US AID, was provided to the Danube countries in transition which had direct or peripheral effect on environmental protection.

25.2.1. HELP TO THE DANUBE PROGRAM

The 1992 Danube Environmental Programme Work Plan (EPDRB, 1992) estimated the cost of Phase I, the initial three year action plan at 48 MECU ($57.6 million) most of which (41 MECU or $49.2 million) was to come from international sources. Although informal commitments totaled about 34 MECU ($40.8 million), a sum this large did not materialize during the initial three-year period.

The EU Phare program and the GEF provided almost all of the support for Phase I. Each had representatives on the staff of the Program Coordination Unit and each was responsible for recruiting these representatives and paying their salaries. The GEF provided the Danube Program with 6.2 MECU ($7.5 million) during Phase I (see Table 25.1). The sum provided the Danube Environment programme by the EU Phare program (MECU 13.4 or $16 million) was insignificant when compared with the total made available to the CEE countries for environmental programs during the first five transition years (see Table 25.2). The Phare program gave almost ten times its Danube Program contribution to individual East European countries for environmental protection. The total sum for all assistance provided to EE countries during the first five years of the transition period was estimated at $100 billion.

Environmental grants were a part of a number of priority technical assistance areas including private sector development, reconstruction and privatization; agricultural and rural development; social development, labor and health, inter alia. Funding under the Tempus (trans-mobility programme for University Studies) program also included some environmentally relevant projects.

25.2.2. US AID

The US AID had received an initial appropriation from the US Congress under the Support for Eastern Europe Democracy (SEED) program which obligated $1.5 billion in aid from 1989 to 1992; of this amount almost half went to Poland. SEED had three primary objectives: to develop market economies and a strong private sector; strengthen democratic institutions to sustain democracy, and improve the basic quality of life in priority geographical areas. US assistance levelled off at about $400 million a year. Originally intended for three countries in Central Europe, it was expected to cover more than a dozen from the Balkans to the Baltic (Hutchings, 1994, p. 188).

25.3. New Phare and GEF Strategies

Funding was requested from the EU Phare and GEF programs for Phase 2, before the transfer of responsibilities to the new Commission not only for operations but to help with the follow-up to the Strategic Action Plan. The requests were the product of a Danube Program Task Force review of basin needs and the revised funding strategies of the two major donors. The following description of specific changes in both funding programs will help to focus on the final decisions that were made to continue funding to the Danube Environment Program.

25.3.1. PHARE PROGRAM

Phare directors reviewed funding strategies in 1994, noted problems, and came up with a plan to take the program to the end of the century. They introduced a standardized and decentralized administrative system to meet CEE needs and promised continued "streamlining" of its procedures. The report concluded that basic policy and institutional foundations had been "largely established" in the CEE partner countries. This permitted Phare to focus on support for "more complex policy, harmonization and awareness needs and in financing the major capital investment projects which are critical if the area is to substantially improve the quality of its environment before the turn of the century (Wilson, 1995, p. 26)."

Environmental Action Plans

Such focus would include help with environmental action plans, and the refinement and harmonization of environmental legislation to bring it in line with that of the EU. Phare would assist with the development of realistic schedules and associated investment programs but expected 90% to come from national sources. The question of environmental liability was to be decided, for example, so that governments would be expected to pay for past damage while privatized enterprises would pay for future clean-up. Clean-up investments would focus on the latter, supporting treatment of past pollution only where this is having a direct impact on human health.

Capital Investment

Capital investment for environment protection is one of Phare's six core target areas for funding. Support for high priority waste treatment equipment for civil projects was emphasized. Effective government policies in place, investments would concentrate on

- alleviation of water pollution, especially in waste water treatment;
- the reduction, at low cost, of damaging air emissions from industrial enterprises, and
- preservation of biodiversity.

Cross-border cooperation is intended to be part of the capital project strategy. Countries would also be helped with the financing of environmental improvements and with intergovernmental coordination, as well as increasing public awareness of environmental issues.

intergovernmental coordination, as well as increasing public awareness of environmental issues.

Total Costs
The estimated total cost of significant improvements in the CEE environment by the year 2010 would be over $300 billion. Donor programs, including Phare, "will contribute probably no more than about seven per cent of total requirements (Wilson, 1995, p. 29)." The EU program reported that assistance during its first five years of operation (1990 to 1994) totaled 4,284 MECU to 11 EU partner countries which made it the largest assistance program of its kind (Wilson, 1995, p. 7). According to the report, "Between 1990 and 1994 Phare financed several hundred environmental projects totally MECU 337 to Albania, Bulgaria, Czech and Slovak Republics, Estonia, Hungary, Latvia, Lithuania, Poland. Romania, and Slovenia." This amounted to approximately 10 per cent of the total Phare budget for this period (Wilson, 1995, p. 9).

25.3.2. GEF STRATEGIES

The Global Environment Facility had been funded for a three-year pilot phase (1991 - 1994) at $1.2 billion to focus on biological diversity, reduction of global warming, protection of international waters, and reduction of ozone layer depletion. Funding during this period was allocated as follows: biodiversity, 48% of the projects and 43% of the funding; global warming, 38% of the projects and 40% of the funding, ozone depletion 2 percent of the projects and 1 percent of the funding, and international waters' protection, 12% of the projects and 16% of the funding. Many aspects of the pilot phase were criticized in an independent evaluation (World Bank, 1994) conducted at the agencies' request. Both the choice of focal areas and the amount of allocations were selected for criticism.

Need for Clear Strategy
The main deficiencies of the international water program, according to the report, were the lack of a clear strategic framework and of well defined priorities within the framework; a definition of international waters that does not recognize that these waters represent a dynamically linked, integrated whole with the waters under national jurisdiction, and inadequate attention to the mismanagement of biological resources. As previously noted the marine environment received 82% of funding; only one river, the Danube, had received funding within the fresh water category. Fresh water related projects, including international river basins, were described as "grossly underfunded," (World Bank, 1994, p. 64).

The GEF developed a new strategy in 1994 as its pilot phase concluded, to provide "a framework for programmatic cohesiveness and integration among the many entities that participate in the GEF, including the three implementing agencies (UNDP, UNEP, and the World Bank), the scientific and technical advisory panel, the GEF Secretariat and the international conventions for which the GEF provides operational supports and funds for implementation (Global Environment Facility, 1996, p. 1)." The operational strategy was designed to achieve a number of goals: to be consistent with national and regional activities, ensure sustainability of global environmental benefits, reduce risk caused by uncertainty, complement traditional development funding, among others.

Pollution Priorities

The strategy emphasized the urgency of pollution in transboundary waters. A consensus had emerged to support a more comprehensive approach to waste resources management, one that is cross-sectoral, integrates ecological and development needs, and is based on holistic analyses of the carrying capacity of the water environment. The GEF role was to be a catalyst to ensure that countries better understand the functioning of their international waters systems, gain an appreciation of how their sectoral activities influence the water environment, and find means for collaborating with neighboring countries to collectively pursue effective solutions.

Priorities were assigned to imminent threats to international waters. These included: control of land-based pollution sources affecting surface and groundwater especially the prevention of releases of persistent toxic substances and heavy metals, as well as reduction of nutrients, biological contaminants, sediments endangering species and ecosystems; and the prevention and control of land degradation, desertification or deforestation; reversing the degradation of critical habitats; control of unsustainable use overfishing, excessive withdrawal of freshwater, resource extraction, and the control of ship-based sources of chemical washings (Global Environment Facility, 1996, pp. 47 - 59).

25.4. Funding Request for Phase II

25.4.1. ACTION IMPLEMENTATION PLAN

The leaders of the Danube Program Task Force submitted proposals for funding Phase II to EC Phare/Tacis and to the GEF in 1996, with an outline of objectives and funding needs. The Task Force had recommended that the Strategic Action Implementation Plan (SIP) focus on transboundary and national tributaries which affect downstream water quality and not on the Danube itself since pollution from tributaries had the greatest impact on the river, the delta and the Black Sea. Capital investment projects to resolve the most urgent transboundary problems were identified but expected to be funded from sources other than GEF or Phare. The requests for grant funding followed a strategy which included a phased expansion of sewerage and municipal waste water treatment capacity, a reduction of discharges from industry, emissions from agriculture, and conservation, and restoration and management of wetland and floodplain areas (EPDRB, 1996).

Grants were requested for the continuation of a coordination unit for the operation of the SIP and activities left over from past agendas. Five broad groups were identified for funding: reduction of contaminants and protection of human health; sustainable land use; wetlands and nature conservation; sustainable use of water resources; institutional capacity building. A sixth group, basin wide projects, included long-term approximation and cooperation actions, public awareness raising, general capacity-building activities, applied research and support to on-going basin projects such as the Accident Emergency Warning System and the monitoring and data sharing activity. A total of 48 projects were proposed under these categories.

25.4.2. GUIDE FOR INVESTMENT

The grant money was to support demonstration projects which would stimulate "sustainable solutions by allocating incremental and domestic (basic) finance both in public and private sectors in the longer term (EPDRB, 1996, p.6)." The amounts available for capital investment from the EIB, World Bank and EBRD are guided by priorities and resource allocation decisions of governments. Direct environmental and related investment by international financial institutions depends on priorities of the borrowing country, the level of debt it can support, the lender's sectoral priorities, the quality of the proposed investment and the financial/managerial strength of the borrower/project owner. Thus the SIP is to be considered as a "stimulus for the major investment pipelines (EPDRB, 1996, p. 6)."

25.4.3. TOTAL SUM REQUESTED

The total sum requested for grants was $89 million. The major portion, $53.3 million, was to be provided by the nine governments in transition, from national resources and/or investment loans or other donor contributions. Requests for GEF funding were set at a probable $12.1 million, with Phare/Tacis estimated at US $23.6 million.

Funding eventually approved by the GEF and anticipated from the Phare Program for the interim transfer period had modified objectives and lower funding levels. It did not include funding for the SIP because the program "does not provide sufficient information on how to establish clear priorities on transboundary concerns, as required by the GEF Operational Strategy (EPDRB, 1996b, p. 6)." The project objectives were to complete 1) the technical phase of priority identification for future investment, 2) the identification of specific priorities, 3) public awareness and participation activities to facilitate such identification, and 4) the development of financing for the Strategic Action Plan. The project was linked with a Black Sea Strategic Action Plan which was to receive a total of $8.8 million with a contribution from the GEF of $1.79 million. It was expected that the first objective would be completed by December 1997 which would permit the formulation of a basinwide program to submit to the IC.

25.5. Global Environment Facility Funding, 1996-1997

The GEF has approved funding for the development of a Danube River Basin Pollution Reduction Program, for a period of 16 months, $3.6 million for the remainder of 1996 and $9.8 million requested for 1997. The availability of parallel funding of $3.9 million for 1996 and $9.8 million for 1997 from the EU Phare programme was anticipated.

25.5.1. OBJECTIVES

The GEF Danube Project has four objectives:

1. Completion of the technical phase for priority identification of future activities with a detailed report on pollution status, including its impact on the Black Sea, amount, location and sources of pollution to be itemized; and an inventory and strategic plan for wetlands, the Danube Delta and floodplains. Also included is funding for a transboundary diagnostic analysis to formulate a basinwide framework to implement identified priority actions.
2. An identification of priorities to facilitate a Black Sea-Basin approach, an endorsement of strategies and programs to foster sustainable institutional and financial arrangements for effective environment management.
3. Public awareness and participatory activities to facilitate priority identification and to increase involvement in targeted public awareness coupled with remedial activities; the development of a coordinated, information and data exchange mechanism which will include internet connections for main data centers, ministries of environment and NGOs.
4. Developing and financing of the Danube Strategic Action Program, including a portfolio of Danube Basin proposed environmental investments, reflecting basin wide consensus and national concurrence for presentation.

A ministerial level conference to review concrete commitments to high priority Danube pollution in 1997 will be followed by a donor-pledging meeting later that year. The proposal to establish a Danube Trust Fund as a mechanism for financial support for the Program will be studied and a framework paper produced to describe the structure, rules, and governance procedures of the mechanism.

25.5.2. FUNDING FOR THE PERMANENT IC

At the first joint meeting of the PCU of the Danube Programme and the Secretariat of the interim IC in July of 1996, problems related to the funding and financing issues which would follow the entry into force of the Danube Convention were reviewed. Costs were of concern as well as possible funding sources. A distinction was made regarding future funding for the IC between its "own" or operational work to be funded by contributions from member countries and from a trust fund, and projects and programs which might be under the guidance of the Commission. The Programme's funders have made it clear that they could not fund "sovereign national tasks" local costs to pay salaries and the costs to run the International Commission's secretariat. A trust fund proposal was to be presented at the 1997 Ministerial Conference. The trust fund could also be used for such activities as applied research projects, updates of basin wide assessments of pollution, inter alia.

Danube Recovery Program

A second proposal, put forward by German representatives suggested a Danube Recovery Programme similar to the European Recovery Program (Marshall Plan) after World War II. The program would be funded mainly by contributions from industrialized countries, especially members of the EU, to fill the gap between national and commercial financing of project investment. A pilot study over a three to five-year period was suggested. The

EU Commission estimates BECU 300 ($4.9 billion) will be needed in the next 15 years to improve the state of the environment in Central and Eastern Europe. The EU Phare program has invested ten percent, MECU 337, of its total grants to Central and Eastern Europe for environmental improvements. An appeal to the EU Conference of Ministers in 1996 was planned.

The idea of instituting a Marshall Plan type program for general aid to CEE had been proposed during the early days of the transition. Ambassador Hutchings (Hutchings, 1994, pp 186-7), an expert in international finance, has pointed out that the Marshall plan was for post-war reconstruction and not related to help for the situation in CEE which he characterizes as severe, sometimes bizarre, "mis-development" which required the opening up of the "inherent potential" of economies and societies. There were lessons in the Marshall Plan experience for CEE during the transition. Regional stability and security needs resembled those in post-war western Europe. Responsibility now, as then, rests with the peoples and governments concerned, and not with international bureaucrats, and, finally, regional cooperation should be a conscious goal of foreign assistance.

25.6. Interim Accomplishments

While uncertainty about its future funding dogged the footsteps of the interim International Commission members and its secretariat, there were markable achievements during the first five years of the Danube Program. The formal Convention, almost certain to go into effect in 1997, was concise and specific about the environmental policies and programs Danube countries were expected to put into place. Danube country legislation would give strength basin-wide to the enforcement of the polluter pays principle. Countries were expected to budget for and raising funds for capital investment for clean-up.

Table 25.1 Danube Program and Related Environmental Funding, 1991-1997/8 ($millions)

For or through EPDRB	1991-1995 Committed	1991-1995 Spent	1996-7/8 Committed	1991-1995. General Funding
CEC PHARE	9.5	13.4	8.1	198.2
EBRD	3.2	4.0	NA	130.4*
GEF	7.1	6.2	3.9	
Netherlands	0.4	0.4	NA	
World Bank	0.4	0.4	NA	
USAID	1.4	6.0	6.4	**
TOTAL	34.0	30.4	18.4	

*EBRD reports that some 30 cities in the Danube Basin will benefit from its environmental funding

**USAID allocated $85 million in all of Eastern Europe and CIS, for 1990-1995

Competition for public funds from funding agencies for capital needs was severe. Approximation of EU laws, a lengthy and complex process. Other issue areas had priority over the environment. Danube country governments, in particular ministries of environment, responded with determination and talent to Danube Program tasks. An evaluation of its achievements and recommendations for its future are in Chapter 26.

TABLE 25.2. Estimated costs, funding proposals for Phase II of Danube Program, 1996-1999 (Amounts in US$millions).

GROUPS AND CLUSTERS	ADDRESSING Phare/Tacis	ADDRESSING GEF GLOBAL ENVIRONMENTAL BENEFIT	COSTS, DONOR SHARING PROPOSAL			
			Baseline	GEF	Phare	TOTAL
CONTAMINANTS AND HUMAN HEALTH	Support to approximation relating to UWWTD and IPPCD; removing constraints of historical degradation	Assist countries to control land-based sources of pollution of water column and Black Sea.	28,395	2,085	4,740	35,220
SUSTAINABLE LAND USE	Supports approximation and raising public awareness; promotes concept of sustainability	Allowing local initiatives to cooperate on transboundary basis to reduce pollution from one of the major user groups	9,545	1,120	1,700	12,365
WETLANDS AND NATURE CONSERVATION	Integrated approach to transboundary wetlands and conservation problems, removing constraints imposed by historical environmental degradation, institutional capacity building	Combined with the two previous Program groups they involve integration of land and water resource management as primary component addressing degradation of international waters. Generally of high global significance	10,835	1,555	1,700	14,090
SUSTAINABLE USE OF WATER RESOURCES	Approximation (directives, standards and institutions associated with river basin management); institutional strengthening; public awareness; implementation of DRPC	International management mechanism to harmonize environmental laws/regulations, and encouraging international collaboration in management of transboundary and shared water	595	1,540	2,625	4,760
BASIN-WIDE PROJECTS Includes public relations, applied research, data and information management, Danube grants program, revisions of SAP, NAPs, capacity building		Further develop capacity existing institutions, experts, assist countries in building common understanding of importance of less polluted water column from source to mouth, assist countries to further develop sustainable environmental and economic policies	0	3,050	8,925	11,975

* Source: EPDRB, Danube Strategic Action Plan Implementation Programme, 1996-1999 ** Baseline = country contribution

CHAPTER 26

Installing an Instrument of Peace: Recommendations

Countries sharing a river basin must agree on how they will jointly use its resources to avoid the harm mis-management by one or a few can bring to all. Specialists in many disciplines can demonstrate optimal use of water resources in each basin country so that all benefit, on an equitable basis. Hydrologists, natural scientists, engineers, economists, and a host of other specialists can and, indeed, in many instances have, met this challenge through models of water availability and use. The quantity and quality of water can be measured and its availability projected. The cause of the deaths of species can be identified and protective steps for their habitats recommended. In the final analysis, however, it takes decisions by governments to agree to put the products of experts to use for their basin.

Conclusions, Recommendations

The intrinsic value of international river basin agreements as instruments of peace in regions prone to conflict has been supported for basins generally and for the Danube especially. The primary goal has been to convert lukewarm agreement support among international and national policy makers to recognition of the crucial role a fully implemented Danube convention can play in the future peace of Europe.

The study has complemented the recent analyses of the ideological, political, and economic transformation of the former eastern bloc countries by demonstrating the relevance of the Danube agreement to transition goals. It sought to fill the gap between such analyses and the extensive work of scientists and technicians who know how to manage the supply and content of basin surface water and other resources but do not know how to stimulate the slow and uneven government and investor response to such needs.

The dismal results of cold-war failures contrasts with the possibilities for a better managed environment in the transition years. Data about national and international pollution levels, the effect of cooperative activities, projected needs as basin countries seek to meet EU criteria, all of these and more, support strong recommendations for changed policies by countries themselves and by the grant and investment agencies.

26.1 PART ONE: Basin Benefits and Challenges

The benefits and challenges of managing international river basins were explored and an optimal model suggested to achieve effective agreements. Summary data about four river basins in developing regions of the world (the Nile and Senegal in Africa; the Mekong in Asia, and the De la Plata in South America) and the Rhine in Europe identify their major conflicts and the level of support for international agreements to resolve them.

26.1.1. CONCLUSIONS, FOUR BASINS

I.1. The need for balanced use of natural resources in each basin is clearly established. The increasing scarcity of water, a high level of population growth, poor farming methods, among other factors, have already created serious hardships, reflected in economic and social indicators. The relationship between resource protection and economic and political stability is also evident.

I.2. Imbalances in resource availability resulting from geographic and hydrogeologic factors have led to political and economic domination by one or more countries over others in basins, contributing to lower standards of living, and crises in international relationships which have occasionally led to violent confrontations. Diversions of streams for the exclusive use of one country and failure to treat biological and chemical wastes are increasingly sources of conflict. Egypt's domination of the Nile, and control of river development in the De la Plata basin by Brazil and Argentina, for example, have fostered imbalances in the sharing of river basin resources.

I.3. The effect of "third parties" on the development of river resources has been a factor in each basin. Colonial pasts have an effect on current investment decisions as commercial and other interests in the former colonial powers continue to dominate their economic development. The environmental effects of hydro-electric power dams and other projects are a matter of debate in all the countries as project planning with inadequate environmental impact assessment continues.

I.4. All of the basins reviewed have had their share of historic national and regional ethnic or tribal rivalries which have, in many instances, escalated to full-scale war. A direct relationship between such conflicts and the use of river resources is in most instances not apparent. It is obvious, however, that this type of conflict interferes with agreements on equitable use of river resources.

I.5. The degree of cooperation among the countries in each of the four basins reflects a general failure to achieve the optimal goals of river basin management. While basin countries have met in common forums to discuss basin issues, some countries continue to be excluded or to exclude themselves from all types of negotiations. Sharing accurate, compatible data and information about all aspects of water availability (quantity and quality), is not practiced in any country although the urgent need for some data, for example for navigation purposes, has prompted a regular exchange. Third parties traditionally do not play an exemplary role with respect to assistance to countries in each basin although Global Environment Facility funding has counteracted criticism of environmental insensitivity.

26.1.2. RECOMMENDATIONS

RI.1. More structured basin studies
Target: International financial institutions and other funding sources

The value of structured studies of international river basins, following the outline used in Part One, should precede support for any particular development project and/or be undertaken periodically during negotiations. Hydrologic and hydraulic engineering data about proposed projects need to be amplified by detailed analyses of the possibilities for and the possible results of long-term agreements. Confrontation over project effects can be avoided if decision processes are reviewed and an appropriate strategy followed.

RI.2. Connect water resource development with policy decisions
Target: Research Institutes

Funding should be made available for applied research projects which explore connections between policy decisions to negotiate the use of river basin resources and the management of water supply and its use. The latter has a solid literature from scientific and technical sources; studies are needed which relate its findings to support for long-term agreements for optimal use of water and other natural resources. Decisions determining the fate of agreements in each of the river basins under discussion in Part One and those elsewhere in the world would be better understood if the rationale of decision-making wase more closely followed. The connection between colonial pasts and current agreements would provide a fruitful area of investigation.

RI.3. Speed up information and data-sharing
Target: NGOs and other actors in the civil society
Those participating in the civil society, however short on resources they might be, have a right to full data and information about the status of the environment in each basin country. International funding institutions should request such information from the countries and provide state-of-the-art assistance to its achievement. Substantiating the factual basis of decisions will serve resource protection, greater public participation in the decisions, and help to avoid future conflicts.

26.1.3. THE RHINE, CONCLUSIONS

I.6. Rhine countries at a high level of agricultural and industrial development faced a critical need to restore and protect shared resources which suffered during the post-

World War II boom. There was no question in the early 1960s that the river desperately needed a common approach to waste treatment A succession of crises, some quite recent, following an initial agreement suggested the need for new approaches.

I.7. Imbalances in the use of the river continue to cause problems. Upstream diversions contributed to severe flooding early in 1995, as many of the heavily populated Rhine cities were submerged. A century of development which had ignored management precautions was responsible. Pollution by upstream countries damaged the downstream waters of the Netherlands.

I.8. The historic rivalries of basin countries presumably have faded or disappeared with membership in the EU. While basin countries have been negotiating conflicts over the use of the river for decades decisions to control pollution have developed slowly. Most of the Rhine countries have adopted or improved on EC Council Directives but there has been a general reluctance to share data. Pressure has increased recently for more data sharing within and among countries.

26.1.4. THE RHINE, RECOMMENDATIONS

RI.4. More focused, available, trend data
Target: European Commission, Europe Environment Agency, EC Phare Program
The publication of data and information about all European river systems is an urgent need. The report, *Europe's Environment*, and its *Statistical Compendium*, contain extensive data about the status of the environment in Europe, including major river basins. These reports are expensive and need annual updating. In most instances the CEE countries lack trend data which will be available in the future. Data and information supplements for these river basins should be published and also made available electronically.

RI.5. Extension of analytical study to other European river systems
Target: International Agencies, Country Governments and Research Institutes
The analytical outline suggested in this study, or some variation of it, should be used for extensive reports on the current status of agreements on international river systems.

26.2. PART TWO: The Basin in Pre-Transition Years

26.2.1. CONCLUSIONS

II.1. Caught in an interplay of competing empires from its earliest recorded history, the Danube region has endured stresses common to international river basins: competition for natural resources, shifting power balances, economic and political development controlled by third parties, and deeply ingrained historic and regional rivalries expressed in violent encounters. Opportunities for countries to support peaceful cooperation for basin resource management were minimal throughout the pre-transition period.

II.2. The varied economic and political development of each east European basin country suggests reasons for their uneven transformation in the transition period. The impact of World War II intensified the economic development begun in pre-war years and chilled nascent democratic movements. The industrial capacities of Czechoslovakia and Hungary produced weaponry and Romania provided agricultural produce and petroleum. Except for Yugoslavia governments were lead by Nazi puppets and vestiges of democratic institutions all but stamped out. War damage to municipal and industrial infrastructures was extensive.

II.3. The Soviet presence in the Danube east European countries from 1945 to 1989 significantly reduced the ability and willingness of basin countries to agree formally or informally on a basin-wide approach to natural resource protection. The free sharing of economic and environmental data and the adoption and enforcement of common water quality standards were not contemplated.

II.4. By 1949 Soviet-style regimes in all the east European countries had versions of the USSR constitution of 1936 in place. The Soviet model rested politically on the "dictatorship of the proletariat" and the public ownership of land and means of production. To an extent, Soviet leaders regarded each country differently, for historic, economic, and political reasons and thus each had its own separate history under communism. In turn, the new Communist leaders within each country manipulated the Soviet government format to suit their tastes, most often with more regard for their own power positions than the needs of the general public.

II.5. Intensive industrial development and wasteful farming methods in the socialist countries during the pre-transition period led to increasing degradation of water and other resources. Environmental legislation which legally established water quality standards, among other environmentally protective goals, was ineffective. There were severe problems meeting World Bank standards for environmental protection that were part of loan agreements for industrial and other projects. Investment programs did not include funding for the construction or renewal of sewage systems and treatment plants.

II.6. Agreement on the shared use of the Danube River Basin remained limited to support for navigation regulations despite suggestions that the mandate of the Danube Commission be broadened to include pollution control. Growing recognition of the danger of transboundary pollution led to a commitment in 1985 by all Danube countries, including Austria and Germany, to monitor the river's water quality.

26.2.2. RECOMMENDATIONS

RII.1. Study the history of the Danube as a regional entity

Target: Research Institutes

Historians should explore the impact of the use of basin resources, eg for transportation, on the events and trends of various periods. Regional studies are appropriately based on other phenomena such as ethnic linkages, economic impacts, and revolutions, but the study of the Danube basin as a region would illuminate relationships between resource use and conflict. Such studies would be valid over time and their findings might be applied to other basins.

26.3. PART THREE: A New Day for the Danube

26.3.1. CONCLUSIONS

III.1. Pressure for a basin-wide Danube agreement which would upgrade environmental policies, including waste treatment, to meet west European standards increased significantly in the 1980s. The Chernobyl disaster had sounded an alarm; east European negligence was a problem for all of western Europe. Western aid post-1989 provided an opportunity for a coordinated rather than piecemeal basin-wide approach.

III.2. By the end of 1991 Danube countries had agreed on specific ways to improve the management of the river basin. International funding agencies suggested the adoption of most of the basic objectives for transboundary river basin management along the lines of those discussed in Part One. Guidance and funding came from the EC as well as the GEF, the EBRD, and the US AID. During 1991, representatives of the Danube countries and the international agencies advanced joint action in the basin by agreeing to negotiate proposals for a formal basin convention and by establishing a framework for cooperative programs. These would guide investment in natural resource protection, monitor water quality and share information about it, and encourage the strengthening of institutions to support these and other basin management goals.

26.3.2. RECOMMENDATIONS

See 26.5. Part Five Conclusions.

26.4. PART FOUR: Profiling Political, Economic and Environment Reforms

26.4.1. CONCLUSIONS

IV.1. The replacement of authoritarian with democratic institutions may or may not be a condition for improved management of river basin resources in other parts of the world; it was an absolute condition for the restoration and long-term protection of the environmental quality of the Danube Basin. Determining the readiness of Danube countries to support joint activities and an international convention requires an estimate of their approximation of western economic and political norms, an understanding of environmental problems, and a more detailed look at changing environmental policies.

IV.2. The first major political reform during the transition, the adoption of new constitutions by most of the countries, was intended to change the functions of each part of the decision process (Figures 1 to 4). Single parties no longer controlled all parts of the process from setting economic plans to endorsing the decisions of a legislature. Newly elected governments represented more than one party and were committed to open the process to a strengthened civil society and to make its products transparent to all.

IV.3. The quality of the transformation to democratic forms of government and market economies in the transition countries has been widely monitored. Six of the Danube countries had signed EU association agreements by 1996 and were being judged on their approximation to EU political and economic goals. A 1996 report evaluates the "state of their readiness for integration" on the basis of the success of their political reform, compatibility of their reformed legal systems with western standards, foreign and security policy orientation, economic reforms and revisions in social policies, and general problems with the multiple requirements of the association process. Table 10.1 records political reform and tables 10.2 to 19.1 reflect changes in social and economic indicators from 1990 to 1995.

IV.4. Political party leaders have not yet acquired a knack for political compromise. Their lack of support from significant numbers of voters means that they tend to represent the interests of small clienteles. Considerable power is concentrated in the hands of members of former communist parties, who tend to block or interfere with the reform process. The proper role of the new civil society is not widely recognized, particularly with respect to nongovernmental organizations. The media are often under government control although new newspapers are appearing in significant numbers.

IV.5. Economic reform has had mixed results in the CEE countries. Current economic indicators, especially national GDPs, need to be evaluated as they compare with those of 1989. Slovenia is the first among the six Danube EU applicants to surpass its 1989 GDP level. Hungary has, since 1989, received by far the highest level of investments from abroad of any CEE country, $10.634 billion, one-third of its GDP. The Czech Republic and, to a lesser extent, Slovakia, have also achieved high economic ratings. Croatia, Romania, Bulgaria, Moldova, and Ukraine have from moderate to severe economic problems.

All of the six applicants for EU membership have stabilized their currencies with the exception of Bulgaria which has experienced significantly high inflation. Structural reforms have had some success; all countries but Bulgaria and Hungary have balanced their budgets. Bulgaria has a critical foreign debt problem. Consolidation and restructuring of the large state-run industrial enterprises is going slowly, limiting foreign and other investment in them. Telecommunications, road and rail networks, continue to need modernization in all countries.

IV.6. While all countries are attempting to prevent the impoverishment of low-income groups, to base health and retirement insurance on user contributions, and to shift such functions to private providers, poverty and poor living conditions continued.

IV.7. Environmental policies have a low priority with government leadership in each country and with the EC as it guides applicants toward EU membership.

IV.8. The basin countries, including candidates for EU membership and those who are not, and Austria and Germany, committed themselves to joint activities in the protection of basin natural resources at the founding meeting of the Danube Environment Program in September 1991. Governments, particularly ministries of environment, have participated in all phases of the Program, including negotiations for and the signing of the Danube Convention in 1994.

IV.9. The Convention itself is a worthy document, concise but ambitious in its goals and specific about the way water resources in the basin should be protected. With proper support it should have a long-range beneficial impact, serving as an instrument of peace as well as a tool for harmonizing environmental policies throughout the basin.

IV.10. Despite lack of trend data it is possible to estimate the status of the environment in each country, noting pollution sources and, in particular, clusters of contaminated sites requiring priority attention to eliminate dangers to public health. Compatible monitoring and laboratory testing techniques are not yet fully in place. A common pattern of pollution in the former eastern bloc countries can be identified which facilitates responsive planning by governments.

IV.11. Each country has passed or amended framework legislation to guide more detailed laws on types of resources and enforcement. By 1994 almost all countries had revised the environmental standards inherited from the Soviet era and developed improved methods of enforcement. Most of the legislation moved slowly through parliaments concerned about what they considered to be more serious matters. Environmental funds for cleanup derived from fines were sometimes targets for corruption; in any case the amount of funding available was small. Some assistance was received from EC Phare and other sources. Western models were widely available even if direct help with relating legislation to the situations in each country was scattered and not as productive as it might have been.

26.4.2. RECOMMENDATIONS

RIV.1. Develop and demonstrate the connection between improved economies and adoption of western environmental norms

Target: Danube Country ministries of environment and major international funders

Decision processes affecting the environmental status of each country need to be accurately understood and demonstrated to government leaders, all ministries affected by the legislation, and to decision makers in the international funding agencies. The connection between environmental policy issues and the economic status of countries must be made so that the "environment" is not placed on a back burner and brought to he fore when it serves the limited ends of political leaders. This is a connection which has been established in western countries through environmental impact assessments, risk analysis, and other techniques that have only recently been discussed in eastern Europe.

RIV.2. Increase and improve access to environmental status data and information
Target: Key actors in the nascent civil societies in each Danube country
Despite cynicism about the speed with which civil societies, as defined by western democracies, are developing in the new democracies of eastern Europe, there are those active in political parties, working for newspapers and journals and for television and radio broadcasting stations, and in public interest groups who understand and support improved environmental policies. Each activist needs access to accurate data and information about all phases of these issues: level of pollutants, the location of heavily polluted sites and related health statistics, effect of present enforcement measures, and the like. The source should be official but findings subject to review by non-governmental experts. This could invigorate languishing environmental movements. Clientele is there: students, teachers, parents in threatened areas, and private sector entrepreneurs with foresight who understand the value of compatibility with western standards.

26.5. PART FIVE: Making and Keeping Long-term Commitments

Each conclusion is followed by a recommendation

26.5.1. CONVENTION

Conclusion

V.1. The cornerstones of an agreement on long-term commitment to protect Danube water resources were in place soon after the end of Soviet domination in the Danube region. Basin countries had committed to a formal Convention, begun to develop and share compatible data on the status of their respective environments, and contributed to a basinwide strategic action plan to restore and protect water and other natural resources. They had accomplished this with guidance and funding from key actors in the donor community.

The Convention points in the right direction but its implementation still seeks funding. Its effectiveness depends on continued cooperation among countries where domestic policy will lead toward fewer restraints on development even while its foreign policy supports integration with the west.

Recommendation

RV.1. Support a broad interpretation of Convention goals
Target: All Danube country governments, funders
There needs to be continuous support for the ultimate goals of the Danube Convention: full cooperation on fundamental water management issues and the undertaking of "all appropriate legal, administrative and technical measures to improve the current

environmental and water quality conditions of the Danube River...and to prevent and reduce as far as possible adverse impacts and changes occurring or likely to be caused."

As the former communist countries embark on full economic recoveries the precedence of joint resource programs goals should be guaranteed by policy makers in their governments and among the international funding agencies if they are to achieve success in national goals..

Conclusion

V.2. The possibility that compliance with western environmental standards will enhance opportunities for economic development aid has motivated ministerial cooperation with the Danube Program and the Convention. Support for the Program has been marginal financially, and weak in other ways. The EC Phare Program continues to lack coordination in its assistance programs as, presumably do other funders such as the GEF and, peripherally, US AID.

Recommendation

RV.2. Connecting Danube basin management with other assistance

Target: EC Phare Program, GEF, US AID

A full report is needed about assistance to the Danube Program and the grants and investments made outside the program, including those tagged for environmental projects or with specific relevance, such as funding for energy-saving projects. The connection between support for the basin-wide program and the compliance of the six applicant basin countries with EU criteria needs to be clearly presented so that there is no question about EC expectations concerning the future of the International Commission.

Permanent staff of the Phare Program, GEF, and US AID rather than consultants should be responsible for implementing coordination of this policy. A basic report each year should identify the total amounts allocated to basin countries through the Danube Program or the successor Commission secretariat and list amounts for environmental projects allocated to each of the eligible Danube countries. Presence or lack of coordination should become more visible, a positive result of this move.

Conclusion

V.3. Direction of the data and information program

While the development of a monitoring network to evaluate water pollution levels is under way, few, if any, of the objectives set for a data and information program by the Danube Program Task Force early in 1992 have been achieved.

There is no access to existing primary information systems and databases developed for the Danube region, its assistance to the country coordinators of the Program is questionable, and information about the multiple Danube Program activities and the status of international and national environmental policies is lacking. The impact of the achievements of the Program has been reduced and public participation in policy issues hampered as a result.

Recommendation
RV.3. Development of and support for information and data access
Target: Danube Program Task Force and IC members; GEF
The NATO May 1996 workshop on access to water quality data and information made a number of specific proposals for adoption by the Danube Program and its successor Commission Secretariat. A follow-up NATO linkage grant will result in a series of electronic products to implement workshop suggestions. Key recommendations from these projects should be discussed and applied.

Conclusion
V.4. Analysis of the political transformation of the Danube basin countries in transition emphasizes the need for a stronger civil society in each of them. In particular, public participation expressed through nongovernmental organizations has failed to gather strength despite special help provided by the Danube Program. Political parties which sprang from politicized environmental movements have lost their identity and their clientele. Many of the organizations which define themselves as NGOs lobbying for improved environmental policies are ephemeral and frequently reflect the authoritarian attitudes of political and cultural heritages. Thus the controversy over the Gabcikovo Dam generated significant funding in Hungary for nationalist reasons. The groups which formed pursued this cause with fervor and for the most part paid little attention to basin-wide issues, or other environmental concerns within their country. As NGOs shifted from dependence on government subsidies (such help continued at a reduced level in most of the formerly socialist countries) to greater independence, help from the west has been poorly coordinated and inadequate.

Recommendation
RV.4. Coordinate financial and technical assistance through information centers in each country
Target: Funding sources for public participation: GEF, EC, inter alia
Since the concept that public participation needs would be fulfilled through the presence of NGO representatives on the Danube Task Force has not proven useful, and information centers in each of two countries have, the latter type of coordination should receive support from Program funders. As originally recommended by a large number of viable NGOs throughout the basin, center staff appointments should be circulated among groups and information and assistance should be freely available to all groups and to the general public. The centers could serve to generate independent fund-raising for NGOs.

Conclusion
V.5. The wide-ranging Strategic Action Plan developed by the Danube Program included some pre-investment studies which yielded data about clean-up needs but went aground due to lack of funding for construction. The supplementary Strategic Action Plan Implementation Program suggested more specific activities to be accomplished by the countries themselves. Tributary basins specifically identified for assistance included the

Sava, Mura/Drava, Morava, Vah, Tisza, Prut/Siret, Jiu and Yantra. Availability of funding seems a remote prospect for the countries with the possible exception of two or three (Hungary, Czech Republic, Slovenia). In question is the continued implementation of commitments made by the countries when the ministers responsible for the state of their environments signed the Danube Convention.

Recommendation

RV.5 Coordinated help with long-term investment in water resource protection

Target: Decision makers in G-24, US AID, private sector investors

Despite the lack of accurate data about environmental clean-up needs a number of reconnaissance and pre-investment studies, complemented by statistics from the countries themselves, suggest estimated totals of around $300 billion for all of CEE. As the Danube Program leaders and IC members grapple with funding needs by facility and by country a figure for basin countries of about $200 billion seems a reasonable estimate.

One urgent task will be the continuous analysis of needs by country, based on economic indicators and environmental trend data as it begins to develop. Priorities may shift from those initially identified and cost-effect measures, recognized by funders, can reduce overall support needs. Energy conservation is an example of a worthy target as a way of reducing costs. The concept of a trust fund to help with Commission programs, particularly within countries, is a good one and appears likely to gather support. The proposal for a Danube Recovery Program similar to the Marshall Plan for the reconstruction of Europe after World War II has two problems. The situation in the Danube region has some similarities to the post-war needs but is not "similar" enough to gather the support of a possible donor country or countries.

26.6. US Policies - a Critique

Most of the recommendations for nourishing the growth of what has been called an instrument of peace for a region prone to violence necessarily have been directed at those immediately responsible for the operation of the Danube Program, its primary funders and the participating countries. The US AID has sent representatives to Task Force meetings, and sponsored a number of projects, including pre-investment studies in four river basins. It also provided initial funding of $5 million to the Regional Environmental Center in Budapest to channel aid to NGOs in each country.

US help has been important as it strengthened the technical capacities of specialists in the East European countries. The high level of east European education in the sciences produced a number of water resource and other experts in the Danube countries who communicated with each other across national boundaries, were trained in west European countries, and followed international developments in their respective disciplines with interest. Help from the United States in the transition years was eagerly sought for environmental and other programs. Some types of assistance were more productive than others. A training course in river basin management in Bulgaria conducted by the Vermont Institute for Sustainable Communities spanned technical and policy issues. It was considered to be effective by Bulgarian local government staff and NGOs. A US AID project for one-on-one development of water quality data systems went well in one country but petered out elsewhere as funding was discontinued.

Advantages that can be gained by the US in providing assistance to the Danube region, especially for its commitment to joint river basin management, are multiple and need not be enumerated here. Coordination is widely recognized as a serious problem for all donor agencies. Uncertainties in budgets is one problem beyond immediate control. The need to turn to outside help when support is lacking for increased staff is a problem for EC Phare, the GEF, and US AID. The results are a diffusion and fragmentation of programs, roadblocks to the development of coherent policies, and, to quote authorities on the EC Phare program, "led to a disturbing bias in actual expenditure away from the intended CEEC beneficiaries." Inevitably consultants received far more money for pre-investment studies than did competitively skilled Danube country specialists.

A recommendation extended to all donors, including US AID, is to submit to an evaluation by objective outside sources of their roles in the Danube Program. It may not be too late, despite the transfer of functions to the IC, for an analysis which plots coordination strategies applicable to all donors and, possibly, to future investment in the basin. After all, western Europe and the United States share the same goals with respect to the long-term value of developing and preserving an instrument of peace in the Danube region.

Table 26.1.Political Transition Danube Countries, 1991-1996

	Bulgaria	Croatia	Czech Republic	Hungary	Romania
CONSTITUTIONS					
Date Adopted	July 1991	December 1990	January 1993	August 1949	December 1991
Rule of Law	Guaranteed, political pluralism, etc		Charter protects many rights, environment	New constitution under review	Strong presidency; no censorship
Human Rights	Protection of human rights, safe environment	Protection of human rights, safe environment	Abides by Human Rights Convention		Abides by Human Rights Convention
HEAD OF STATE					
Current	Petar STOYANOV	Franjo TUDJMAN	Vaclav HAVEL	Arpad GONCZ	Emil CONSTANTANESCU
Date Elected	November 1996	May 1991	January 1993	August 1990	November 1996
Elected by	Popular vote	Popular vote	Parliament	Parliament	Popular vote
LEGISLATURE					
Prime Minister	Undecided 1997	Nikica VALENTIC	Vaclav KLAUS	Gyula HORN	Victor CIORBEA
Date Elected	January 1995	April 1993	June 1992/1996	July 1994	December 1996
No. of Houses	One	Two	Two	One	Two
Elections when	NA 1997	NA 1997		Spring 1998	
Coalition in power	Bulgarian Socialist Party-led coalition; 125 of 240 seats	Upper & Lower Houses	Civic Democratic Party-led coalition; 65 of 200 seats	Socialist Party; 209 of 386 seats	
WESTERN ORIENTATION					
EU Agreement	YES	NO	YES	YES	YES
NATO Partner	YES	NO	YES	YES	YES
UN Member	YES	NO	YES	YES	YES

Table 26.1 continued

	Slovak Republic	Slovenia	Moldova	Ukraine
CONSTITUTIONS				
Date Adopted	January 1993	December 1991	July 1994	Using 1978; new one being drafted
Rule of Law	Broad definitions of national property	Includes	Provides for multi-party government, et al	July 1994
Human Rights	Human Rights Convention	Human Rights Convention		
HEAD OF STATE				
Current	Michal KOVAC	Milan KUCAN	Petru LUCINSCHI	Leonid D. KUCHMA
Date Elected	February 1993	December 1992	December 1996	July 1994
Elected by	Parliament	Popular vote	Popular vote	Popular vote
Term		5 years		
LEGISLATURE				
Prime Minister	Vladimir MECIAR	In Doubt	Andrei SANGHELI	Pavlo LAZARENKO
Date Elected	December 1994	January 1997	December 1996	
No. of Houses	One	Two	One	One
Elections when	July 1992	December 1996	1999	NA
Coalition	Movement for a Democratic Slovakia heads coalition;	Liberal Democracy, United List of Social Democrats, et al	Agrarian-Democratic Party	Check text for information about parties
Western Orientation				
EU Agreement	YES	YES		
NATO Partner	YES	YES	YES	YES
UN Member	YES	YES	YES	YES

Table 26.2. Optimal International River Basin Agreement

Negotiating	Formal Convention	Sharing Data	Development Plan	Third Parties, Investors et al
All countries	Signed by all basin countries; All basin provides for commission receiving funding support and delegated decision authority	State-of-the-art monitoring, recording and sharing of water quality data, all procedures internationally compatible; other data shared and available to public according to agreement	Countries agree to provide environmental impact assessments for all major industrial, agricultural, water resource projects; resource protection goals in basin-wide plan	Investors (international funders, private sector) abide by development plan consult with international commission at earliest stages and throughout; invite public scrutiny

Table 26.3. Status of International River Basin Agreements, Four Basins

Basin	Negotiating	Formal Convention	Sharing Data	Development Plan	Third Parties
Nile	Among all countries to a very limited degree; Technical Cooperating Committee has trained technicians	No basin-wide; 1959 Nile Waters Agreement to allocate waters bet. Sudan and Egypt in effect	Sudan and Egypt share data to verify agreement, otherwise limited, at discretion Sudan, Egypt	No basin-wide plan despite growing water scarcity; basin use dominated by Egypt	Post-Aswan no test for environmental sensitivity available; US aid sustains Egyptian economy
Senegal	Four countries (minus Guinea) agreed early after independence to discuss reaction to development plan	Convention produced an international association, the OMVS; potential basin-wide management agreement	Takes place to a limited degree but levels of accuracy and compatibility are uncertain	Under discussion among the countries; a basin wide plan of a limited nature agreed to but has not yet funded	The basin countries have resist international development plans, insisting on resource protection
Mekong	Common forum for discussion and negotiation has been limited to lower Mekong, excludes China, Myanmar	Four countries 1957 agreement with international committee, now called Commission; Secretariat implements decisions	Some increase in hydrologic monitoring stations but monitoring and sharing of environmental data is limited	Development plan accepted by four countries in 1995; data not sufficient for pro-environment groups	Despite UNDP promises to protect environment no environmental impact assessment for 100 dams
De la Plata	Forum for negotiations from time to time; some bi- and tri-lateral agreements for large-scale navigation and hydroelectric projects	Two basin-wide agreements , one establishing the CIC for joint development project and one to establish a common market in the basin.	Data shared under very limited conditions: for projects and navigation	Size and complexity of river systems makes planning for resource development and protection difficult	Concern for environmental damage widespread but no environmental impact assessment in use

Table 26.4. Status of International River Basin Agreements, the Rhine, 1995

Negotiating	Formal Convention	Sharing Data	Development Plan	Third Parties
Discussion in place since 1950s; productive negotiations following crises	Formed international commission (ICPR) with administrative functions but much discretion in hands countries	Data being shared increasingly in response to need but on voluntary basis; some public sharing of data in response to NGO pressure	No plan by multilateral agreement attempted; Rhine Action Plan may be a force for greater concern about water resource development	Greater sensitivity to results of investment in projects without environmental impact assessment; EU participation an important factor

Table 26.5. Status of International River Basins Agreement, the Danube, 1989

Negotiating	Formal Convention	Sharing Data	Development Plan	Third Parties, Investors
Limited to navigation, some tributary basins, ineffective monitoring	Navigation Agreement implemented by Danube Commission	Water quality monitoring at 11 sites on Danube; delay in reporting; methodology not calibrated	Not attempted	Funding of multiple dam projects mostly in Austria and Germany questioned by environmental groups; Gabcikovo project an international problem

Table 26.5.. Status of International River Basin Agreement, the Danube, 1996

Negotiating	Formal Convention	Sharing Data	Development Plan	Third Parties, Investors et al
All countries except Serbia discuss agreement on joint decisions at technical, ministerial level	Danube River Protection Convention signed at ministerial level, ratification and entry into force anticipated 1997	Much data on environmental status of natural resources has been shared; but automated networks for accurate, compatible environmental status in progress	Basin-wide plan limited to identify priority needs for waste treatment programs; basin-wide impact assessments of industrial, agricultural enterprise not yet undertaken	Investors in transition sensitive to environmental concerns; provided guidance for agreement

Glossary

AEWS	Accident Emergency Warning System
BAT	Best Available Techniques
BEP	Best Environmental Practice
CEE	Central and Eastern Europe
CIS	Commonwealth of Independent States
CMEA	Council for Mutual Economic Assistance
CPC	(Danube) Country Program Coordinator
CSCE	Conference on Security and Cooperation in Europe
DRPC	Danube River Protection Convention
EAP	Environmental Action Plan
EBRD	European Bank for Reconstruction and Development
EC	European Commission
ECU	European Currency Unit
EEA	European Environment Agency
EIB	European Investment Bank
EPA	Environmental Protection Agency
EPDRB	Environmental Program for the Danube River Basin
EU	European Union
GEF	Global Environment Facility
IBRD	International Bank for Reconstruction and Development
IC	International Commission (for the Protection of the Danube River)
ICPDR	International Commission for the Protection of the Danube River
ICPR	International Commission for the Protection of the Rhine
ISO	International Standards Organization
IUCN	World Conservation Union
m	meter
km	kilometer
MECU	Million European Currency Units
MLIM	(Danube Program) Monitoring, Laboratory, and Information Management
NAP	National Action Program
NAS	National Academy of Sciences (United States)
NATO	North Atlantic Treaty Organization
NGO	Non-Governmental Organization
NRA	National Rivers Authority (United Kingdom)
OECD	Organization for Economic Cooperation and Development
PCU	(Danube) Program Coordination Unit
Phare	EU Program of assistance for economies of central and eastern Europe
PMTF	(Danube) Program Management Task Force
RAP	Rhine Action Program

REC	Regional Environmental Center, Budapest
sq km	square kilometer
TNMN	Danube Program Trans-National Monitoring Network
UNDP	United Nations Development Program
UNECE	United Nations Economic Commission for Europe
UNEP	United Nations Environment Program
USAID	United States Agency for International Development
USEPA	United States Agency for Environmental Protection
VITUKI	Hungarian Water Resources Research Institute
WWF	World Wide Fund for Nature
WWTP	Waste water treatment plant(s)

SELECTED BIBLIOGRAPHY

Balek, Jaroslav. (1996). "Transboundary Diagnostic Analysis of the Environmental Concerns in the Danube River Basin," Global Environmental Facility, June 1996

Bandi, Gyula and Stanislaw Wajda. (1996). "Regional Overview," in Regional Environmental Center, *Approximation of European Union Environmental Legislation*, Budapest: Regional Environmental Center.

Bicanic, Ivo. (1994) "The Croatian Economy: Transition and Stabilization," in U.S. Congress, Joint Economic Committee Report, *East Central European Economies in Transition*, Washington, DC: GPO.

Biswas, A. K.. (1992). "Sustainable Water Development: a Global Perspective," *Water International* 17 (1992):68-80.

Brada, Josef C. (1994). "The Slovak Economy after One Year of Independence," in U.S. Congress, Joint Economic Committee Report (1994), pp.518-530.

Braun, Marc. (1997). "The Rhine: monitoring programmes and exchange of data within the ICPR," in Murphy (1997), pp.23-30.

Bricke, Dieter, et al (1995). "Slovakia," in Weidenfeld (1995), pp. 199-224.

Brilly, Mitja, et al (1997) "Crossborder Monitoring of Water Quality in Slovenia" in Murphy (1997), p. 189-198.

Brown, J. F. (1988). *Eastern Europe and Communist Rule*, Durham, NC: Duke University Press.

Brusis, Martin, and Cornelius Ochman. (1996), in Weidenfeld (1996), pp. 9-28.

Bryson, Bill (1992). "Linking Europe's Waterways," *National Geographic,* August 1992.

Buksa, Zeljko. (1995) "The Impact of War," *Danube Watch*, 1:3, 4.

Burelli, Marco. (1992). "Nile delta sinking," *Alternatives* 18 (1992):4

Butora, Martin. (1995). Lecture, Woodrow Wilson Center, East European Studies, Washington, DC, April 1995

Carothers, Thomas, (1996). *Assessing Democracy Assistance: the Case of Romania*, Washington, DC: Carnegie Endowment for International Peace, 1996.

Cano, Guillermo J. (1987). "The Del Plata Basin: Summary Chronicle of its development process and related conflicts" in Vlachos (1987)

Chomchai, Prachoom. (1987). "The Mekong Committee: An Exercise in Regional Cooperation to develop the lower Mekong Basin" in Vlachos (1987)

Chorus, Ingrid. (1994). "A Healthy Danube Environment: Developing the Environment and Health Dimension of the Strategic Action Plan for the Danube River Basin," unedited draft of a final WHO-UNDP report, made available by the Danube Environment Program.

Hungarian Ministry of Environment. (1994). "Danube Basin Ecological Convention, fourth draft," Budapest, unp.

Davies, Norman. (1996). *Europe: a History*, Oxford: Oxford University Press.

Deimel, Johanna and Rumen Dimitrov. (1996). "Bulgaria," in Weidenfeld (1996), pp. 20-50.

Deininger, Rolf A. (1987). "The Survival of Father Rhine," *AWWA Journal*, July, 1987.

Dimitrov, Rumen. (1995). "Bulgaria," in Weidenfeld (1995).

Dinu, Adriana and Matthew Rowntree. (1995). "Moldova," in Philip R. Pryde. (1995), pp. 193-203.

Dupont, Christophe. (1993). "The Rhine: A Study of Inland Water Negotiations," in Gunnar Sjostedt (1993). *International Environmental Negotiation*, Newbury Park, CA: Sage. p. 135-148.

Durkee, Linda C. (1989). "Risk Communication and the Rhine River," *International Environment Reporter*, Bureau of National Affairs, 12:10, Oct 11, 1989.

EBRD (European Bank for Reconstruction and Development), *Transition reports and updates*, London: EBRD, 1994, 1995, 1996.

EC. Council Regulation No. 1210/90, *Official Journal of the European Communities*, No L 120/I, 11/5/90.

EC. (1995). White Paper: "Preparation of the Associated Countries of Central and Eastern Europe for Integration into the Internal Market of the Union," Brussels: March 5, 1995.

EC.(1994). "European Union Relations with the Countries of Central and Eastern Europe," Background Brief BB19 (updated January 1994).

Economist, April 20th, 1996; May 4, 1996; June 8, 1996; October 5, 1996, November 2, 1996 and, November 23, 1996

EPDRB (Environmental Programme for the Danube River Basin). (1992). Programme Work Plan, 14 February 1992, 1992 (distributed at Danube Program Task Force Meeting, unp.)

__. (1993). *Annual Report, September 1993,* Brussels, Belgium.

__. (1995). *Annual Report 1995*, Vienna: 1995.

__. (1995a). *The Strategic Plan for the Danube River Basin, 1995-2005*, Vienna.

__. (1996). "Danube Strategic Action Plan Implementation Programme, 1996-99," draft 4 July 1996, Vienna.

__. (1996a) "International Commission for the Protection of the Danube River," Vienna.

__. (1996b). Programme Coordination Unit, Danube River Program, "Draft Summary Report, 1st Joint Meeting of the Interim International Commission for the Protection of the Danube River and the Task Force for the Environmental Programme for the Danube River Basin," July 25-26, 1996, Vienna, Austria.

__. (1996c). "Proposal for Review," unp, provided by the GEF Coordination, UNDP New York, October 21, 1996.

Erdey, Gyorgy and Mariann Karcza. (1995). "Hungary," in Regional Environmental Center (1995).

Eurostat et al. (1995) eds., Luxembourg:Eurostat et al, 1995.

Falkenmark, Malin and Gunnar Lindh, (1993) "Water and economic development," in Gleick (1993).

Financial Times, 2 March 1994.

Fisher, Sharon. (1996). "Slovak Parliament Approves Controversial Law on Foundations," OMRI Analytical Brief, 1:191, June 21, 1996.

Fitzmaurice, John. (1996). *Damming the Danube: Gabcikovo and Post-Communist Politics in Europe*, Boulder, CO: Westview Press.

Freeman, William E. (1995). "Environmental Management in Ukraine," in Pryde, Philip R. (1995). ed, *Environmental Resources and Constraints in the Former Soviet Republics*, Boulder: Westview Press, pp. 159-173.

Gasparikova, Bozena, et al. (1996). "Slovak Republic," in Regional Environmental Center (1996).

Gleick, Peter H. (1993) ed, *Water in Crisis*, New York: Oxford University Press.

__ (1993a) "Water and Conflict," *International Security*, 18:1 (Summer 1993). pp. 79-112.

Glinda, Jozef and Gabriela Fischerova (1995). "Slovak Republic," Regional Environmental Center (1995).

Global Environment Facility, "Operational Strategy," Washington, DC, February 1996

Green, Peter S. (1996). "Analysts say Reform in Czech Republic Will Survive Election," *International Herald Tribune*, June 4, 1996

__. (1996a). "Czech Bank ailure Linked to Securities Deals," *International Herald Tribune*, September 21-22, 1996

Habcu, Simon et al, "Romania," in Regional Environmental Center (1996).

Halasz, Nicholas (1959). *In the Shadow of Russia*, Ronald Press, New York.

Hammond, Allen and others. (1995). "Environmental Indicators: A systematic approach to measuring and reporting on environmental policy performance in the context of sustainable development," Washington, DC, World Resources Institute, May 1995.

Handl, Vladimir, et al. (1995). "Czech Republic," in Weidenfeld (1995).

Hedges, Chris, "In Croatia's Capital, Politics and Democracy Don't Mix Well," *The New York Times,* May 2, 1996

__, "Croatia Chief Seeks to Muzzle a Radio that Aided His Rise," *The New York Times*, October 9, 1996

Herrnfeld, Hans-Holger. (1996). *European by Law: Legal reform and approximation of law in the Visegrad countries*, Gutersloh: Bertelsmann Foundation Publishers

Hertzman, Clyde et al. (1995). *Environment and Health in Central and Eastern Europe, A report for the Environmental Action Programme for Central and Eastern Europe*, Washington, D.C.:World Bank.

Hinrichsen, Don and Gyorgy Enyedi, eds. (1990). *State of the Hungarian Environment, Budapest: Hungarian Academy of Sciences*

Hock, Bela and Laszlo Somlyody (1990). "Freshwater Resources and Water Quality", in Hinrichsen (1990).

Hungarian delegation to the Danube Ecological Conference, 12-13 February 1991, "Statement," Budapest: unp.

Hori, Hiroshi, "Development of the Mekong River Basin, Its Problems and Future Prospects," *Water International*, 18 (1993).

Hupkes, R. (1990). "Pollution of the North Sea imposed by West European rivers (1984-1987)", Amsterdam: International Center of Water Studies.

Hutchings, Robert L. (1994). "Five Years After: Reflections on the post-Communist Transition and Western Assistance Strategies," in U. S. Congress, Joint Economic Committee, pp.176-190.

Inotai, Andras, and Jurgen Notzold (1995). "Hungary," in Weidenfeld (1995).
International Herald Tribune, October 8-9,1994, 1,5.
__, September 25, 1996, 15.
International Rivers Network, "Hidrovia Campaign, Dossiers 1," Berkeley, CA: March 1995, and 2, October 1995
IUCN. (1992). Environmental Status Reports: vol. 4 Conservation Status of the Danube Delta, Norwich, UK: Page Brothers.
IUCN (World Conservation Union) (1994). *Analysis and Synthesis of National Reviews for the Danube River Basin Environmental Programme*, prepared by Edward Maltby, et al, Gland, Switzerland: IUCN.
Jancar-Webster, Barbara (1991). "The East European Environmental Movement and the Transformation of East European Society," in Jancar-Webster, ed, *Environmental Action in Eastern Europe*, Armonk, N. Y.: M. E. Sharpe. pp. 192-219
Jennewein, Marga and Kristina Larischova. (1996), "Czech Republic," in Weidenfeld (1996), pp 51-72.
Kaiser, Phillip J. (1994). "The Czech Republic: An Assessment of the Transition," in US Congress, Joint Economic Committee Report (1994), pp. 506-517.
Kinzer, Stephen, "West Says Slovakia Falls Short of Democracy," *New York Times*, Dec. 26, 1995.
Kiss, Alexander, "The Protection of the Rhine Against Pollution," *Natural Resources Journal*, 25:3 (July 1985).
Klinda, Jozef and Gabriela Fischerova, "Slovak Republic," in Regional Environmental Center (1995).
Kocikova, Pavla, "Czech Republic," in Regional Environmental Center (1996).
Koulov, Boian. (1995). "The Democratization Process in Bulgaria: Parliamentary Elections and Political Parties," unpublished paper.
Kuntchev, Rumen, lecture at Woodrow Wilson Center, 25 January 1995.
Lack, Tim. (1997). "The Work of the Monitoring, Laboratory and Information Management Sub-Group," in Murphy (1997) pp. 127-134.
Lesnic, Mihai. (1995). "Romania" in Regional Environmental Center (1995).
Lammers, Owen. (1995). "A New River Development Policy for UNDP?" *World Rivers Review*, 10:2, 1995.
Lammers, Owen, et al. (1994). "Considering the Hidrovia: A Preliminary Report on the Status of the Proposed Paraguay/Parana Water Way Project," International Rivers Network Working Paper 3, June 1994.
LeMarquand, David G. "International Development of the Senegal River," in Vlachos (1987).
Linnerooth-Bayer, Joanne and Susan Murcott. (1995). "The Danube River Basin: International Cooperation for Sustainable Development," unpublished draft, October 1995.
Marer, Paul. (1994). "Hungary during 1988-1994: a political economy assessment," US Congress, Joint Committee Report, pp. 480-505.
Marsh, Virginia (1996). *Financial Times*, April 6-8, 1996.
Mastrovic, Margita, (1995) "Croatia," in Regional Environmental Center, (1995).

McCaffrey, Stephen C., (1993) "Water, Politics, and international law," in Gleick (1993).

Mencinger, Joze, and Arnold Suppan, "Slovenia,"(1995(in Weidenfeld, (1995), pp.225-250.

Mindjov, Kliment, "Bulgaria," (1995) in Regional Environmental Center (1995).

Mnatsakanian, Ruben A. (1992). *Environmental Legacy of the Former Soviet Republics*, Edinburgh: Center for Human Ecology, Edinburgh.

Mohammed, Azez Ali. (1994). "The Role of International Financial Institutions," US Congress, Joint Economic Committee, pp. 191-204.

National Environmental Action Plan, 1995, Chisinau, Moldova, unp.

Morehead, Alan (1960). *The White Nile*, New York: Dell Publishing

Murphy, Irene L[yons]. (1987)"The workshop in retrospect: an interpretive viewpoint," in Vlachos (1987). Washington, D. C.

__, INFODANUBE (1993). An Ecological Information System for the Danube RIver Basin, on diskette, Budapest: Regional Environmental Center.

__, (1996). "Environmental Policies in Transition: A NATO partnership country symposium," in Murphy, 1996

__, ed.(1997). *Protecting Danube Resources: Ensuring Access to Water Quality Data and Information*, Dordrecht, Netherlands: Kluwer.

Murphy, Irene L. and J. Eleonora Sabadell. (1986). "International River Basins: A policy model for conflict resolution," *Resources Policy*, 12:1 (June 1986): pp. 133-144.

National Geographic (1984), *Encyclopedia of Rivers*, Washington, DC.1984

OECD, *Environmental Policies in Yugoslavia*, Paris: OECD, 1986

____, Water Resource Management: Integrated policies, Paris: 1989

OMVS (1995). (Organization pour la mise en valeur du fleuve Senegal) Magazine "L'Agence de gestion des ouvrages communs," No.2, February 2, 1995.

OMRI (Open Media Research Institute Publications), Daily Digest II, No. 62, 27 March 1996.

Paranici, Bogdan O. and Ioana Petrescu. (1994). *Report of the Workshop on Public Participation and Information Strategies in Danube Basin Countries*, Annex 2, "Networking, Information Exchange and Issue Action Initiatives in the Danube Basin, Sinaia, Romania, February 1994.

Pearce, Fred. (1993). "Dammed to destruction," *The Independent on Sunday*, 6 June 1993, 52, with references to Stanley, Daniel and Andrew Warne article in *Science*, April 1993.

__, "Greenprint for rescuing the Rhine," *New Science*, 26 June 1993.

Pehe, Jiri. (1995). lecture, Woodrow Wilson Center, April 5, 1995.

Perlez, Jane, *New York Times*, December 25, 1994.

__, "With old tricks, Romania's old reds curb press," *New York Times*, June 8, 1995.

__, "Romania bans currency deals by foreign banks," *New York Times*, March 23, 1996.

__ "Despite U.S. Hope and Help, Ukraine's Star Fades," *New York Times*, June 27, 1996, A3-J.

__, "A Bulgarian is Slain and Motives are Many," *New York Times*, November 3, 1996.

__, "Looted by its own Officials, Bulgaria Faces the Day of Economic Reckoning," *New York Times*, October 28, 1996.

__, "Incumbent in Romania, a Leftist, Faces Runoff," New York Times, November 5, 1996.

Pesic, Vesna. (1996). *Serbian Nationalism and the Origins of the Yugoslav Crisis*, Washington, DC: United States Institute of Peace

Pomazi, Istvan and Andras R. Csanady. (1995) "Hungary," Regional Environmental Center. (1995).

Pomfret, John, "Dissent Mounts in Serbia, Croatia," *Washington Post*, November 23, 1996.

Postel, Sandra, "Water and Agriculture," in Gleick (1993).

Pry, Robert H. (1993). Foreword, *International Environment Negotiation*, ed by Gunnar Sjostedt (1993). Newbury Park, California: Sage Publications.

Pryde, Philip R., ed, *Environmental Resources and Constraints in the Former Soviet Republics,* Westview Press, Boulder, 1995.

Rakonczay, Zoltan, "Wildlife Protection and Nature Conservation," in Hinrichsen, *State of the Hungarian Environment Budapest*, 1990.

Regional Environmental Center (1995). *Status of National Environmental Action Programs in Central and Eastern Europe*, Budapest, Hungary.

__ (1996). *Approximation of European Union Environmental Legislation*, Budapest, Hungary.

Roncerel, Annie. (1996). UNDP. NY. interview October 21, 1996

Roneak, Peter, et al. (1996). "Data Sharing Between the Slovak and Czech Republic in the Morava River Basin," in Murphy, 1996

Rothert, Steve, (1995). "Lessons Unlearned: Damming the Mekong River," I International Rivers Network, Working Paper 6, October 1995.

Rupert, James, "Moldovan Reformer Now Blocks Change, " *Washington Post*, May 10, 1996.

Sbragia, Alberta (1996). "Environmental Policy: the 'Push-Pull' of Policy Making," in Helen Wallace and William Wallace, *Policy-Making in the European Union*, Oxford: Oxford University Press,

Shalash, S. (1983). "Degradation of the River Nile," *Water Power & Dam Construction*, July 1983

Simonian, Haig (1995). "Floods of tears on the Rhine," *Financial Times*, February 8, 1995

Sjostedt, Gunnar. (1993). ed., *International Environment Negotiation*, Newbury Park, California: Sage.

Staar, Richard F. (1988). *Communist Regimes in Eastern Europe*, Stanford, California: Hoover Institution Press.

Spitkova, Timea (1995). "Privatization the key to halting chemical pollution," *Danube Watch,* 1(1995):4.

Stanners, David and Philippe Bourdeau (1995) *Europe's Environment: the Dobris Assessment,* Europe Environment Agency, Copenhagen: Office for Official Publications of the European Commission.

Stephenson, Nigel (1996). Global Markets, *Washington Post*, April 14, 1996
Stiblar, Franjo D. (1994). "Slovenia: A Country Study," US Congress, Joint Economic Committee Report, pp. 599-621.
Ullman, Richard H. (1996). ed., *The World and Yugoslavia's Wars*, New York: Council on Foreign Relations.
UN Food and Agriculture Organization (1984). *Systematic Index of international water resources treaties, declarations, acts and cases, by basin*, Vol. II, Rome.
UNDP, memo from Yves de San, Chief, Division for Europe, 5 May 1991.
UNDP. (1994). *Workshop on Drinking Water Related Environmental Health Aspects in the Catchment Area of the River Danube, Bratislava, the Slovak Republic, 16-19 December 1993,* Vol. II, Country Reports, New York: UNDP.
UNECE, Environmental Conventions, New York: UN ECE, 1992
U.S. Congress (1994). Joint Economic Committee Report, *East Central European Economies in Transition*, Washington, DC: GPO
U.S. Department of State (1994). SEED Act Implementation Report: Fiscal Year 1993, Washington, DC: US Department of State.
Varkonyi, Tibor and Gyozo Kiss. (1990). "Air Quality and Pollution Control," in Hinrichsen (1990).
Vlachos, Evan (1993). "Incentives for Regional Cooperation with Regard to Shared Water Resources," in Proceedings, UNEP-UNDP Freshwater Consultative Forum, Geneva, 13-16 Dec 1993.
Vlachos et al. (1987) eds. *The Management of International River Basin Conflicts: Proceedings of a Workshop Held at the International Institute of Applied Systems Analysis*, Washington, D. C.: The George Washington University.
Vukovich, Gyorgy (1990). "Trends in Economic and Urban Development and Their Environmental Implications," Hinrichsen (1990).
Wagner, Vida Ogorelec. (1995). "Slovenia," Regional Environmental Center (1995).
Weidenfeld, Werner. (1995) ed. *Central and Eastern Europe on the Way into the European Union*, Gutersloh: Bertelsmann Foundation Publishers
__, (1996). ed, *Central and Eastern Europe on the Way into the European Union,* Gutersloh: Bertelsmann Foundation Publishers.
Whittington, Dale and Elizabeth McClelland. (1992). "Opportunities for Regional and International Cooperation in the Nile Basin," *Water International*, Vol 17 1992.
Wilson, Julian, et al, (1995). "Progress and strategy paper: Environment to the year 2000," European Commission, Brussels.
World Bank. (1995). Monitoring Environmental Progress: A Report on Work in Progress, Washington, DC: World Bank.
__, (1994), *Global Environment Facility: independent evaluation of the pilot phase*, World Bank: Washington, DC.
World Bank, "Hungary: Structural Reforms for Sustainable Growth, A Country Study," Washington, DC., 1995
World Rivers Review (1993). Berkeley, CA.

WRc. (1993). "Environmental Standards and Legislation in Western and Eastern Europe towards Harmonization, Final Report Task IV (b): Water Sector Case Study, Part I," London: European Bank for Reconstruction and Development and Commission of the European Communities.

Wyzan, Michael L. (1994). "Bulgaria: A Country Study," in US Congress, Joint Economic Committee Report, pp. 531-551.

Zimmermann, Warren. (1995). "The Last Ambassador: A Memoir of the Collapse of Yugoslavia," *Foreign Affairs*, March/April 1995.

* * * * *

Author's note:

English has been the common language in the Danube Basin since the beginning of the transition. Sources for this book have included works which draw extensively on the other languages of the Basin. Help with data and information about all parts of this book has been provided by specialists in each of the countries.

Data in tables comes from a variety of sources. Sources developing and reporting on economic indicators frequently differ, both in their selection of indices and their content. Thus it is difficult to cite sources for the tables presented. Many of the indicators were checked with the most recent publication to include 1995 indicators, *International Marketing and Statistics, 1997.*

Index

The GeoJournal Library

1. B. Currey and G. Hugo (eds.): *Famine as Geographical Phenomenon.* 1984 ISBN 90-277-1762-1
2. S.H.U. Bowie, F.R.S. and I. Thornton (eds.): *Environmental Geochemistry and Health.* Report of the Royal Society's British National Committee for Problems of the Environment. 1985 ISBN 90-277-1879-2
3. L.A. Kosiński and K.M. Elahi (eds.): *Population Redistribution and Development in South Asia.* 1985 ISBN 90-277-1938-1
4. Y. Gradus (ed.): *Desert Development.* Man and Technology in Sparselands. 1985 ISBN 90-277-2043-6
5. F.J. Calzonetti and B.D. Solomon (eds.): *Geographical Dimensions of Energy.* 1985 ISBN 90-277-2061-4
6. J. Lundqvist, U. Lohm and M. Falkenmark (eds.): *Strategies for River Basin Management.* Environmental Integration of Land and Water in River Basin. 1985 ISBN 90-277-2111-4
7. A. Rogers and F.J. Willekens (eds.): *Migration and Settlement.* A Multiregional Comparative Study. 1986 ISBN 90-277-2119-X
8. R. Laulajainen: *Spatial Strategies in Retailing.* 1987 ISBN 90-277-2595-0
9. T.H. Lee, H.R. Linden, D.A. Dreyfus and T. Vasko (eds.): *The Methane Age.* 1988 ISBN 90-277-2745-7
10. H.J. Walker (ed.): *Artificial Structures and Shorelines.* 1988 ISBN 90-277-2746-5
11. A. Kellerman: *Time, Space, and Society.* Geographical Societal Perspectives. 1989 ISBN 0-7923-0123-4
12. P. Fabbri (ed.): *Recreational Uses of Coastal Areas.* A Research Project of the Commission on the Coastal Environment, International Geographical Union. 1990 ISBN 0-7923-0279-6
13. L.M. Brush, M.G. Wolman and Huang Bing-Wei (eds.): *Taming the Yellow River: Silt and Floods.* Proceedings of a Bilateral Seminar on Problems in the Lower Reaches of the Yellow River, China. 1989 ISBN 0-7923-0416-0
14. J. Stillwell and H.J. Scholten (eds.): *Contemporary Research in Population Geography.* A Comparison of the United Kingdom and the Netherlands. 1990 ISBN 0-7923-0431-4
15. M.S. Kenzer (ed.): *Applied Geography.* Issues, Questions, and Concerns. 1989 ISBN 0-7923-0438-1
16. D. Nir: *Region as a Socio-environmental System.* An Introduction to a Systemic Regional Geography. 1990 ISBN 0-7923-0516-7
17. H.J. Scholten and J.C.H. Stillwell (eds.): *Geographical Information Systems for Urban and Regional Planning.* 1990 ISBN 0-7923-0793-3
18. F.M. Brouwer, A.J. Thomas and M.J. Chadwick (eds.): *Land Use Changes in Europe.* Processes of Change, Environmental Transformations and Future Patterns. 1991 ISBN 0-7923-1099-3
19. C.J. Campbell: *The Golden Century of Oil 1950–2050.* The Depletion of a Resource. 1991 ISBN 0-7923-1442-5
20. F.M. Dieleman and S. Musterd (eds.): *The Randstad: A Research and Policy Laboratory.* 1992 ISBN 0-7923-1649-5
21. V.I. Ilyichev and V.V. Anikiev (eds.): *Oceanic and Anthropogenic Controls of Life in the Pacific Ocean.* 1992 ISBN 0-7923-1854-4

The GeoJournal Library

22. A.K. Dutt and F.J. Costa (eds.): *Perspectives on Planning and Urban Development in Belgium.* 1992 ISBN 0-7923-1885-4
23. J. Portugali: *Implicate Relations.* Society and Space in the Israeli-Palestinian Conflict. 1993 ISBN 0-7923-1886-2
24. M.J.C. de Lepper, H.J. Scholten and R.M. Stern (eds.): *The Added Value of Geographical Information Systems in Public and Environmental Health.* 1995 ISBN 0-7923-1887-0
25. J.P. Dorian, P.A. Minakir and V.T. Borisovich (eds.): *CIS Energy and Minerals Development.* Prospects, Problems and Opportunities for International Cooperation. 1993 ISBN 0-7923-2323-8
26. P.P. Wong (ed.): *Tourism vs Environment: The Case for Coastal Areas.* 1993 ISBN 0-7923-2404-8
27. G.B. Benko and U. Strohmayer (eds.): *Geography, History and Social Sciences.* 1995 ISBN 0-7923-2543-5
28. A. Faludi and A. der Valk: *Rule and Order. Dutch Planning Doctrine in the Twentieth Century.* 1994 ISBN 0-7923-2619-9
29. B.C. Hewitson and R.G. Crane (eds.): *Neural Nets: Applications in Geography.* 1994 ISBN 0-7923-2746-2
30. A.K. Dutt, F.J. Costa, S. Aggarwal and A.G. Noble (eds.): *The Asian City: Processes of Development, Characteristics and Planning.* 1994 ISBN 0-7923-3135-4
31. R. Laulajainen and H.A. Stafford: *Corporate Geography.* Business Location Principles and Cases. 1995 ISBN 0-7923-3326-8
32. J. Portugali (ed.): *The Construction of Cognitive Maps.* 1996 ISBN 0-7923-3949-5
33. E. Biagini: *Northern Ireland and Beyond.* Social and Geographical Issues. 1996 ISBN 0-7923-4046-9
34. A.K. Dutt (ed.): *Southeast Asia: A Ten Nation Region.* 1996 ISBN 0-7923-4171-6
35. J. Settele, C. Margules, P. Poschlod and K. Henle (eds.): *Species Survival in Fragmented Landscapes.* 1996 ISBN 0-7923-4239-9
36. M. Yoshino, M. Domrös, A. Douguédroit, J. Paszynski and L.D. Nkemdirim (eds.): *Climates and Societies – A Climatological Perspective.* A Contribution on Global Change and Related Problems Prepared by the Commission on Climatology of the International Geographical Union. 1997 ISBN 0-7923-4324-7
37. D. Borri, A. Khakee and C. Lacirignola (eds.): *Evaluating Theory-Practice and Urban-Rural Interplay in Planning.* 1997 ISBN 0-7923-4326-3
38. J.A.A. Jones, C. Liu, M-K. Woo and H-T. Kung (eds.): *Regional Hydrological Response to Climate Change.* 1996 ISBN 0-7923-4329-8
39. R. Lloyd: *Spatial Cognition.* Geographic Environments. 1997 ISBN 0-7923-4375-1
40. I. Lyons Murphy: *The Danube: A River Basin in Transition.* 1997 ISBN 0-7923-4558-4

KLUWER ACADEMIC PUBLISHERS – DORDRECHT / BOSTON / LONDON

MIX
Papier aus verantwortungsvollen Quellen
Paper from responsible sources
FSC® C105338

If you have any concerns about our products,
you can contact us on
ProductSafety@springernature.com

In case Publisher is established outside the EU,
the EU authorized representative is:
Springer Nature Customer Service Center GmbH
Europaplatz 3, 69115 Heidelberg, Germany

Printed by Libri Plureos GmbH
in Hamburg, Germany